The Great British Pier

WITH PHOTOGRAPHS FROM
THE FRANCIS FRITH COLLECTION

COLWYN BAY, THE PIER PAVILION 1900 46268

The Great British Pier

WITH PHOTOGRAPHS FROM
THE FRANCIS FRITH COLLECTION

And a Victorian Guidebook Commentary to the Resorts

First published in the United Kingdom in 2006 by The Francis Frith Collection for Bounty Books
a division of Octopus Publishing Group,
2-4 Heron Quays, London E14 4JP, England

Hardback edition ISBN 10: 0-7537-1441-8
ISBN 13: 978-0-7537-1441-6

British Library Cataloguing in Publication Data

The Great British Pier

The Francis Frith Collection
Frith's Barn, Teffont,
Salisbury, Wiltshire SP3 5QP
Tel: +44 (0) 1722 716 376
Email: info@francisfrith.co.uk
www.francisfrith.com

Designed and assembled by Terence Sackett

Printed in Singapore by Imago

Front Cover: Eastbourne, The Pier 1925 *77946t*

The colour-tinting in this book is for illustrative purposes only, and is not intended to be historically accurate.

Every attempt has been made to contact copyright holders of illustrative material.
We will be happy to give full acknowledgement in future editions for any items not credited.
Any information should be directed to The Francis Frith Collection.

AS WITH ANY HISTORICAL DATABASE, THE FRANCIS FRITH ARCHIVE IS CONSTANTLY BEING CORRECTED AND IMPROVED,
AND THE PUBLISHERS WOULD WELCOME INFORMATION ON OMISSIONS OR INACCURACIES

Contents

Piers - An Introduction

THERE CAN BE very few people living in the United Kingdom who do not have fond memories of going to the seaside. As youngsters we love to play on the beach, whilst as adults, our first glimpse of the sea is often sufficient to bring back those childish urges. And what structure typifies a coastal resort more than a pier? How exciting it is to have the opportunity of walking out over the water without getting your feet wet!

The history of the seaside pier can be traced to the growth of towns like Scarborough and Margate in the late 1700s. It became the fashionable thing for the rich to go to the coast, where they could take the natural waters, as distinct from the waters at the mineral springs at inland spas – Bath is the prime example of such a spa. People strolled on jetties, which started to have extra functions as well as being merely for promenading.

Jetties, as early piers tended to be called, were also built for a more practical purpose - somewhere that boats could call. Inland transport was not anything to write home about; at this period, original Roman roads were still being used in many places. Thus it made sense for passengers to get to these new resorts by sea, and to land at a freshly-constructed jetty in the absence of an artificial or natural harbour.

Ryde Pier, built in 1813-4, was the earliest of the ninety or so seaside piers which were to be erected over the next hundred years. More famous, however, was Brighton Chain Pier, built in 1822-3 in the style of a suspension bridge. It attracted the patronage of William IV, and appeared in paintings by both J M W Turner and John Constable. With facilities which included a camera obscura, shower baths, kiosks and shops, the pier certainly had a true pleasure function.

The pier craze soon began to gather momentum, and in the 1830s other English piers opened at Southend, Walton, Herne Bay and Deal. Wales got its first pier in 1846 at Beaumaris, whilst even in Scotland a number of piers had been built before the end of the 1840s. These, including Beaumaris, tended to be more simple constructions, erected solely for ships to land.

Any resort of note felt that it needed a pier to enhance its status. During the peak decades for pier building in England and Wales, the 1860s and 1870s, an average of almost two structures opened each year. Eugenius Birch was the most renowned of a generation of pier engineers: he was responsible for designing no less than fourteen individual constructions. Brighton West was his undoubted masterpiece, and dates from 1866. There was a slight decline in the years that followed, but nevertheless piers opened at a rate of more than one every year right up until Fleetwood saw the light of day in 1910.

Not surprisingly, the British seaside was by then attracting a much wider clientele than just the wealthy. The increasingly powerful middle classes got into the habit of going to the seaside, as did the working classes once they were able to do so. A series of Factory Acts from 1833 onwards gradually reduced the length of the working week, while Parliament introduced four Bank Holidays under legislation of 1871. Hastings Pier opened on the first August Bank Holiday in 1872, with Cleethorpes Pier following suit a year later. In the industrial north, annual summer breaks became widespread. Whole towns would decamp en masse to the seaside – Skegness was known as 'Nottingham-by-the-Sea' owing to the influx of outside visitors.

Yet were it not for the expanded railway network, many seaside towns would have remained coastal backwaters. Track mileage grew from less than 2,000 miles in 1840 to over 15,500 miles in 1870. Passengers were carried at speeds previously undreamed of: averages in excess of 40 mph were by no means uncommon. Again, the Government helped, passing a law in 1844 which compelled all companies to run at least one covered train - some early carriages were little more than cattle trucks - in each direction at no more than a penny a mile. The principle of travel for all was established.

It is perhaps worth noting that many pier promoters were connected with the railways. That was clearly true of Skegness, and also of Cleethorpes, where the pier soon came under the control of the Manchester, Sheffield and Lincolnshire Railway.

Why did pier building come to a conclusion? It was not due to the First World War, but largely because all places that wanted piers already had them. Indeed, some towns, such as Southsea and Great Yarmouth, boasted two. Blackpool, however, reigned supreme with three, and might have had a further two. Yet even Blackpool must doff her cap to Atlantic City, New Jersey, where eight piers have existed over the years. This, incidentally, helps to demolish the myth that no piers were built overseas.

British purists, however, like to regard their piers as being somewhat superior to their American counterparts, which – so they say – lack the grace one associates with a traditional iron-built pier. Whether they are right or not, some of the last piers constructed in England were among the finest, and cost a lot of money. Weston-super-Mare's second pier, the Grand, involved expenditure of £120,000, and Brighton's Palace Pier cost even more at £137,000. Given price levels at the turn of the century, it is not surprising that this remained a record. For these vast sums, you certainly gained a substantial pier. Weston-super-Mare's Grand Pier had a vast pavilion, measuring 150ft by 90ft, which was capable of drawing crowds of 2,000 to events ranging from opera to music hall. Brighton Palace Pier's pavilion contained a 1,500-seater theatre, along with smoking and reading rooms.

Though new pier construction drew to a close (Weymouth Bandstand, which opened 1939, was the last of all), that did not prevent individual piers from continuing to develop in order to provide holidaymakers with the best possible entertainment. The inter-war period was arguably the 'golden age' for piers, with people wishing to put the horrors of conflict behind them. Clacton Pier was at the forefront of this amusement era when it erected its famous funfair. A dance hall, casino, swimming pool and children's theatre were among the new attractions provided. The late Lord Delfont (who was to become a major pier owner in the 1980s and 1990s through his company First Leisure) started his theatrical career as a dancer on this particular pier.

Elsewhere, new pavilions were erected at Southport, Penarth, Morecambe Central, Sandown, Colwyn Bay and Weston-super-Mare Grand amongst others. Shows and orchestral concerts were a popular feature of the successful pier - most of the famous names in music and show business were to perform on piers, from Sir Malcolm Sargent and Sir Henry Wood (of 'proms' fame) to Vera Lynn and David Nixon. And let us not forget George Formby, who played at St Anne's. His name is associated with piers through the song 'The Wigan Boat Express', which is about the supposed mythical structure at Wigan. Piers, you see, are not meant to be found inland!

Ironically, however, there had been a pier of sorts at Wigan: Bankes Pier, a coal tippler on the Leeds and Liverpool Canal. Demolished in 1929, its stone base remained. Today, the entire site is covered by a tourist centre known as 'Wigan Pier' – Wigan appears to have had the last laugh after all! Yet the boats associated with piers in their heyday did not carry coal, but people. Thousands of holiday-makers were brought to the piers by paddle steamers with evocative names like the 'Medway Queen' and 'La Marguerite'. Current trips by PS 'Waverley' help continue this long tradition.

So though economic depression affected certain areas, British piers gained from the affluence many others enjoyed. Those in employment had a shorter working week, paid holidays, and were gradually becoming car owners. Hence the boom years looked as if they could go on for ever. Alas, they did not.

The Second World War was to affect piers in various ways. Naval forces took over both Southend and Bognor Regis, along with Cowes Victoria on the Isle of Wight. Plymouth Hoe and Southsea Clarence were bomb casualties, whilst Minehead was pulled down to give guns a clear line of sight. This also caused the demolition of Deal's second pier, which had been already wrecked by a ship collision. Many piers were breached – that is, a section of decking

8

was removed as a precaution against enemy invasion. However, it seems more likely that a sophisticated enemy would have chosen to land at a secluded bay, miles away from a crowded built-up town like Great Yarmouth, Brighton or Bournemouth. Even if piers escaped sectioning, they remained closed and at the mercy of what the weather had to offer. Steamer services ceased, with pier railways running for the last time at Felixstowe and Herne Bay.

Running repairs, always essential for structures so vulnerable, took a back seat during war-time. Thus when the guns finally stopped firing in 1945, most piers needed a great deal of work to be done before they could open their gates to the public once again. Let us remember that these were still very much years of austerity, with rationing and shortages the order of the day. Given the economic environment, piers could be seen as an unnecessary luxury. So it was that some piers which survived the war did not last that long after it. Six piers disappeared from our coastlines in the 1950s, including Seaview Chain Pier, the first pier to be listed. At Cleethorpes they declined to re-attach the long pier neck; the iron was sold instead to help build a new stand for Leicester City Football Club. Along with Felixstowe, it was a shadow of its former self.

However, there were piers that thrived at that time: 5,750,000 people went on Southend Pier in the 1949-50 season, an all-time record. Deal got itself a new replacement pier, which was opened in 1957 by the Duke of Edinburgh. The new Southsea Clarence Pier followed four years later, a structure unique in being wider than its length. Changing leisure patterns, however, were hitting British resorts hard. No longer were holiday-makers content with their week or fortnight at home; they were wanting to go overseas, where sunny weather was almost guaranteed. Cheap package holidays on sale in Macmillan's 'You've never had it so good' Britain meant that they could afford to do this. Increased car ownership led trippers to move from town to town, affecting smaller resorts particularly. Obviously, piers suffered through a down-turn in visitor numbers – a further eight were demolished by 1980.

Yet by then the tide had started to turn. People were realising the architectural importance of seaside piers, and what would be lost if nothing were done. Groups were set up to try and save the structures at Southend, Clevedon and Brighton (West), supported by a National Piers Society. Bangor became the first preserved pier in 1988, followed by Clevedon and Swanage. The launch of the National Lottery helped the latter two, which both qualified for funding. Other piers also received money from this source.

Meanwhile, there remained the commercial successes, including the three piers at Blackpool. If piers adapted, their prime site meant they could thrive in a competitive leisure market. They were helped by the fight-back of Britain as a holiday destination, with visitors on short breaks and foreign tourists helping to maintain their viability. So though the 1990s saw Morecambe (Central), Shanklin and Ventnor pulled down, the outlook is by no means bleak. We can feel confident that piers can look forward to many years of history.

The text contained within dotted borders accompanying these photographs is extracted from a late Victorian tourist guidebook to the seaside resorts of Britain. It is therefore genuine and offers a true picture of the preoccupations and attitudes of its period. Resorts and individual features are thus referred to in the present tense, and the original spelling and punctuation have been retained for the purposes of authenticity.

The Victorian commentaries themselves are hardly nostalgic, for like a modern guidebook they offer sensible and practical advice to visitors. They describe the type of holiday they might expect from each resort – the quality of the beach, whether it was sand or shingle, how safe was the bathing, whether children were well catered for, what walks there were in the neighbourhood and what local beauty spots were to be recommended.

For us, the buildings we see in the photographs are imbued with nostalgia; we cannot help viewing them as period architecture. Yet at the time the photographs were taken, in many of the buildings the mortar had barely dried. Time and again the Victorian writer tells of local councils and joint-stock companies embarking on radical development along the seafronts, and building lodging-houses, hotels, dining-rooms and a plethora of places of amusement.

We must remember, too, that just 50 years earlier than the date of the earliest of these photographs most of the resorts pictured in this book barely existed. Previously there had been just a few fishermen's huts and an empty seafront where a few hardy travellers whiled away the hours and took rooms at the local inn. Close to the turn of the 20th century, the Victorian writer still refers to them by the sedate title of 'watering-places'.

The words and pictures in this book inevitably recall an age that is long since gone. Yet they bring us a strong reminder, in our age of the ever more exotic foreign holiday, of the simple and diverse pleasures that the old-style British resort with its pier and promenade offered to many generations of tourists. Today we may choose to travel to distant continents rather than take our holidays in our own country, but it is hoped that this book will help bring back happy memories of the British seaside resort and pier in its heyday, that so many of us enjoyed.

ABERDOUR

ABERDOUR, THE STONE PIER 1900   45912

In addition to this old stone pier, there was a more modern construction, seen here on the right. The breezy east coast of Scotland has fewer piers than the west coast, largely because artificial harbours had already been built before the pier-building age. Both of the piers seen here served as landing places.

The sea-bathing village of Aberdour and the old castle are situated three miles westward of Burntisland, and may be reached from the latter place by a pleasant shore path through the woods, which have been considerably damaged by the construction of the railway to North Queensferry. This is a favourite seaside resort of 'trippers' from Edinburgh, steamers plying almost hourly from Leith in summer. The railway from Edinburgh to Burntisland runs via the Forth Bridge and Aberdour.

REDCAR, THE PIER 1896 37594T

REDCAR

Visitors are unanimous in praise of the fine air, broad sands, and picturesque cliffs and ravines. This little town lies a few miles south from the mouth of the Tees. A sea wall has been constructed, forming a fine terrace 30 feet wide, and there are two promenade piers. The sands of Redcar can nowhere be surpassed in extent, being ten miles in length and a mile broad at low water. They have been characterized as 'smooth as velvet, yet so firm that neither horse nor man leave their imprint on them as they tread the strand'. Redcar consists mainly of one long street, with the backs of the houses on one side turned to the sea. As a watering-place, Redcar dates from 1842, since when it has progressed at quite an extraordinary rate, and certainly owes much to its magnificent sweep of sands.

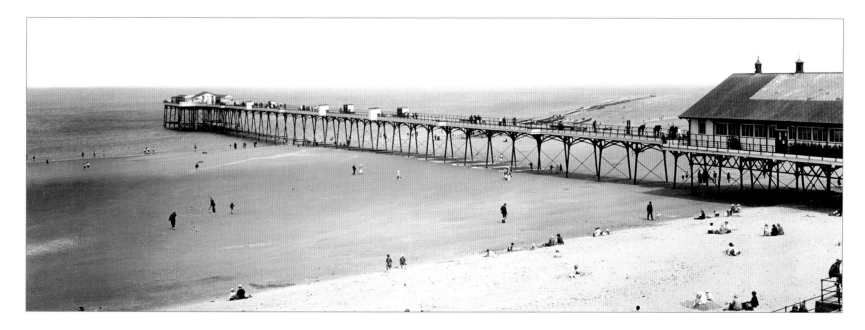

The pier at Redcar opened in 1873. It was constructed by Head Wrightson, an engineering firm from nearby Stockton-on-Tees with a national reputation. Pleasure cruises used to run from a small landing stage, but a colliding ship wrecked this in 1885. Robert Conway's tariff list (shown on the wall of the kisok in 37594t (right) is typical of late 19th-century advertising signs.

There had been a 700-seater bandstand at the pierhead, which was burnt down during an 1898 fire. After a mobile bandstand on the promenade had proved inadequate for local entertainment needs, it was decided to construct a new pavilion, this time at the shore end. It opened in 1907, complete with a ballroom. The other entrance buildings remained the same.

The pier originally measured 1,300ft, but was a victim of sectioning during the Second World War. Mine and storm damage meant that the long neck was removed, leaving Redcar with a pier that measured just 45ft beyond the pavilion. Eventually this too was demolished, in 1980–81.

Opposite: REDCAR, THE BEACH C1955 R16052

Above: REDCAR, THE PIER 1923 74245

Right: REDCAR, THE PIER 1896 37594T

'Sands ... smooth as velvet, yet so firm that neither horse nor man leave their imprint on them as they tread the strand...'

SALTBURN-BY-THE-SEA

This is the only pier remaining within Yorkshire's traditional boundaries. It was opened in 1869, and was to suffer over the years from a series of storms. These reduced its length from an initial 1,250 feet. A cliff lift takes passengers to the pier's entrance.

In the 1970s, it seemed that Saltburn could lose its pier – the local authority actually applied for listed building consent to demolish it on safety grounds. But a public enquiry recommended that only the 13 end trestles should be removed. Today's 681ft-long pier has the same 1920s entrance building, which now contains amusements rather than a theatre.

Opposite: SALTBURN-BY-THE-SEA, THE PIER ENTRANCE 1913   66354

Top: SALTBURN-BY-THE-SEA, THE PIER 1913   66358

Above: SALTBURN-BY-THE-SEA, THE CLIFF TRAMWAY AND THE PIER C1955
S51099

This watering-place, situate about eight miles south of the Tees mouth, in the bay named after that river, is the property of Sir J W Pease and partners, who have taken a very great interest in developing the delightful features of the place as a seaside resort. Saltburn-by-the-Sea is accessible from all parts, whether on the North-Eastern, Midland, or London and North-Western Railway Companies' systems.

The town stands about 130 feet above the sea-level, and when the visitor stands on the promenade, which passes along the top of the sea banks, he has a splendid view over the North Sea; while on the other side of the Tees Bay may be seen the old seaport of Hartlepool, with its lighthouse on a projecting headland at the northern extremity of the bay. A band plays twice daily, either on the pier or in the gardens. Facing the terraces on the high ground there is a promenade, laid out with walks and seats. The ground on which Saltburn stands is broken on the east side of the town by a valley, through which flows the Skelton Beck.

This valley is beautifully laid out as an ornamental pleasure garden, and is unique of its kind at any watering-place.

The elegant iron-girder bridge across the valley is 136 feet high, 25 feet wide, and 600 feet in length, and gives access to the country beyond the gardens. Opposite the pier an inclined tramway conveys visitors to and from the promenade.

WHITBY

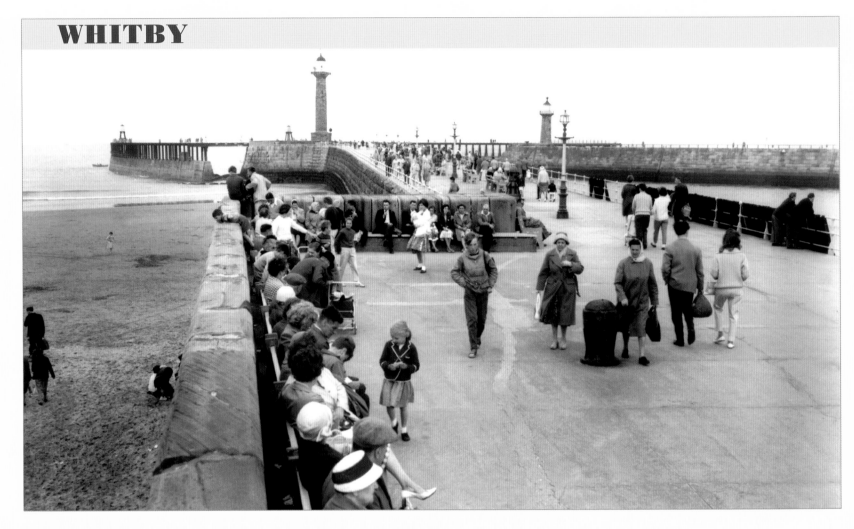

Is this a true pier? It is left out of most reference books, but perhaps it should not be, for it is popular with promenaders, and provides a point of call for boats. It is complete with seats and lights, and only the pedant could completely ignore its claims.

Whitby harbour was renowned for its narrow entry between the two piers with their lighthouses. The whaling ships had to wait for the high tides to carry them through safely. The little fishing cobles had no problems, provided the wind was in the right quarter. There was great rejoicing when the new extensions were finally erected, having first been suggested by William Scoresby a hundred years before. The competent engineers left a space for turbulent seas to break through and lessen the strain on the masonry.

Above: WHITBY, THE PIER C1955 W81152

Opposite above: WHITBY, THE NEW PIER EXTENSIONS 1913 66265A

Opposite below: WHITBY, THE LOWER HARBOUR 1891 28854

The unspoilt gem of this coast

If Scarborough be the queen of watering-places in the north of England, there are many who find her court too gay, and pronounce Whitby the unspoilt gem of this coast. It lies between two cliffs at the mouth of the romantic River Esk, looking out upon the German Ocean, and, like so many other watering-places, writes, as it were, much of its history in the visibly broad line of demarcation that marks new town and old. The latter is on the eastern cliff, where the grey ruins of the noble abbey still attract the eye. This cliff is 250 feet high, and so steep that the houses seem literally huddled together, tier resting upon tier, and mass upon mass. The rest of the town, New Whitby, rises in more dignified aspect on the western cliff, surmounting its very top, and exhibiting conspicuously its splendid hotel. The two towns are united by a stone bridge of three arches, with a movable centre-piece, allowing the passage of ships to the inner harbour.

A peculiar feature of Whitby is the Museum, which contains thousands of fossils, for which the neighbourhood is celebrated. Among these is a gigantic crocodile, 18 feet long. There are many beautiful walks, especially along the cliffs, which rise 600 feet above the sea. The number of persons employed in the jet manufacture is 500.

SCARBOROUGH

SCARBOROUGH, NORTH BAY AND THE PIER 1891 28825

Scarborough's pier was begun in 1866 and opened three years later. The construction costs were £15,000, and it was masterminded by the most famous of pier architects, Eugenius Birch. It suffered damage when a ship collided with the supports in 1883. An entertainments pavilion was added at the shore end at the end of the 1880s to entice further visitors. However, the structure was finally washed away by fierce storms in the winter of 1905. Unfortunately, it was not insured, and so was never rebuilt.

Spread out like an amphitheatre upon a bay and promontory, its houses rise tier behind tier away from the sea. The season here lasts from May to October, and during the greater part of this time the fashionable South Cliff, with its terraces, walks, and handsome music-hall, is crowded with pleasure-seekers from all parts of the kingdom. The bathing at Scarborough is famous. Uncontaminated by any large river, the open bay provides water of the greatest purity, transparency, and saltness; the sand is clear, firm and smooth. The scene on the sands on a fine morning is extremely animated. While some visitors are gambolling among the waves, others are riding along the sands on donkeys or horses. Charmingly-dressed ladies may be seen sitting on the rocks, reading, sketching, or engaged in ostensibly useful 'work'; the old castle, the pier and harbour, the brick houses of the old town, and the handsome range of buildings on the cliff forming a beautiful background to the view.

HULL

Now called 'The Marina', the Humber is still a busy working area. Hull's main industry was deep-sea fishing, regrettably now in decline due to over-fishing by most European countries. Approaching the waterside is the picturesque 'old town' where historic buildings nestle cheek by jowl with street cafes lining the cobbled streets of Hull's traditional heartland where the River Hull joins the Humber.

Above: HULL, THE HUMBER 1903   49820

Of course the most important feature of the town is the docks. The Hull river itself forms a natural dock – narrow, but thronged with vessels and lined with warehouses for a distance of a mile and a half. The Humber Dock was opened in 1809, and the Prince's Dock in 1829. The former dock communicates with the Humber by a basin, protected by piers. The passage across the docks is maintained by means of drawbridges.

CLEETHORPES

This pier was built between 1872 and 1875, funded by the Cleethorpes Promenade Pier Company to the cost of £10,000. The shareholders went on to lease it to the Manchester, Sheffield and Lincolnshire Railway Company (like many piers it was slow to cover its costs, and local contributors were often notoriously reluctant to wait for very long for a return on their investment). Later, the railway company bought the pier outright, keen to capitalise on the huge number of visitors arriving on their trains. The concert hall at the seaward end was a great attraction, but burnt down in 1903. The structure suffered damage during the Second World War, and the seaward section was demolished, so reducing the pier from its original length of 1,200ft to 330ft.

This is rapidly becoming the most crowded watering-place in Lincolnshire. It is quite a unique development of railway enterprise, belonging, as it does, almost entirely to the Manchester, Sheffield and Lincolnshire Railway, who have built a sea-wall and promenade, laid out public gardens, and constructed a good pier. Cleethorpes, owing to its easy railway access, is invaded daily in the summer by enormous crowds of excursionists from Yorkshire, Lancashire and the Midland counties. On this account the pleasant little town is scarcely to be recommended to resident visitors. The sands and the bathing here are inferior to those at Skegness, but there is far more life in the place.

Left: CLEETHORPES, THE BEACH AND PIER C1955   C112131
Above: CLEETHORPES, THE PIER 1906   55740

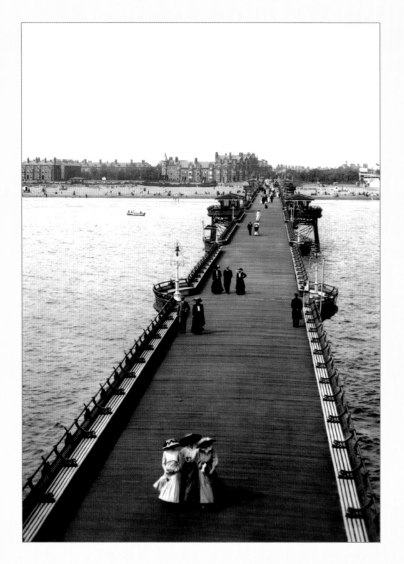

East Coast piers had to go out a long way in order to reach the sea at all tides. Skegness was certainly no exception, and opened in 1881 with a length of 1,817ft. A small notice advertises R Connell, one of the many divers who made a living jumping off the end of piers (44350).

To encourage people to walk along a pier, attractions were invariably provided at the head. Those at Skegness included a 700-seater saloon cum concert hall, which was extended in 1898 with the addition of new refreshment rooms. Steamboat trips ran from a landing stage.

Modernisation of the entrance took place in the late 1930s. The entrance then incorporated a café and shops either side of an archway. However, they were later demolished to make way for a new large entrance building, containing amusement arcades. This opened in 1971.

A severe storm in January 1978 washed away two large sections of pier, leaving the theatre isolated at the seaward end. Rebuilding plans came to nothing, and the ruined pierhead and decking were eventually demolished. However, the section of pier passing over the central beach was extended in the 1990s.

The flat coast of Lincolnshire is not very well off for watering-places. Skegness sprung up into considerable note since the extension of the railway in 1873. The great attraction is the firm wide sands, on which donkeys, swings, cocoanut-shies, and other amusements for excursionists will be found in full activity during the season. There is an iron promenade pier with pavilion. Dancing and concerts take place in another pavilion in the Pleasure Gardens. From the sand hills along the shore there are extensive views over the German Ocean. Fine sunsets may be watched hence. The curious optical illusion known as the mirage is often seen here to perfection during fine weather, when the sea has the appearance of a sheet of glass. During the dark nights of summer, the phosphorescence of the sea is a very charming sight. At such times, as one walks along the shore by the side of the receding tide, each footprint glows with phosphorescent light.

SKEGNESS

Opposite left:
SKEGNESS, FROM THE
PIER 1910   62842

Opposite right:
SKEGNESS, THE PIER
ENTRANCE C1955   S134078

Right:
SKEGNESS, THE PIER 1899
44350

Below:
SKEGNESS, THE BEACH
AND THE PIER C1960
S134150

HUNSTANTON

Left:
HUNSTANTON,
THE GREEN AND THE
PIER 1927   79723

Below:
HUNSTANTON,
THE PIER C1950
H135072P

Opposite above:
HUNSTANTON,
THE FUN FAIR C1955
H135117

Opposite below:
HUNSTANTON,
THE PIER SKATING
RINK C1965   H135115

A desperate leap

Just before the War in 1939, two women were sitting enjoying the sun at the end of the pier when they saw smoke and suddenly found themselves cut off from the shore by a fire that had broken out in the pavilion. Desperate and terrified, they had no choice but to leap into the sea thirty feet below, where they thrashed around in the waves until they were rescued by firemen. As a result of the conflagration, the cafe and concert hall were destroyed, as well as the musical instruments of a visiting concert band.

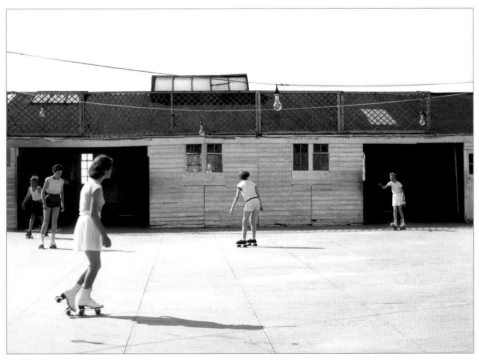

Hunstanton Pier opened on Easter Sunday 1870 with a length of 830ft. Paddle steamers ran across the Wash to Skegness pier a year after the latter structure was built. A pavilion, described as being 'handsome and commodious' in a 1907 guide book, was added later.

The ironwork on this pier was particularly outstanding; structures like this help to show just why piers are regarded as important engineering constructions. At one stage, before the First World War, concerts were held in the pavilion every morning, afternoon and evening, indicating the resort's popularity.

Sad to say, a fire on 11 June 1939 destroyed the pavilion, which was never replaced. After 1945 the pier was used by roller-skaters, and it had a small zoo. The Ealing Comedy 'Barnacle Bill' (called 'All at Sea' for the American market) starring Sir Alec Guinness was filmed here in 1956. A waxwork exhibition and bingo were among the many other post-war attractions here. The entrance building seen in photograph H135072p was replaced by a two-storey construction, which opened in 1964. This survived the storm of 1978 that destroyed the pier, but was burnt down in 2002.

On the north-west angle of the coast of Norfolk stands the pretty watering-place of Hunstanton St Edmunds, which, during the summer months, is crowded with visitors, the rooms, which out of the season can be got for five shillings fetching a guinea a week, or more. This latter fact is not surprising when we consider that the little town is perched upon a hill 60 feet or 80 feet above the sea-level, the top of which is a chalk down; the western side forms a picturesque sea-cliff, overlooking a pleasant and safe beach which extends far seaward at low water. It is the only watering-place on the east coast of England with a western aspect, and it commands extensive views of the opposite coast of Lincoln, 20 miles across.

HUNSTANTON, THE PIER 1921 71033P

CROMER

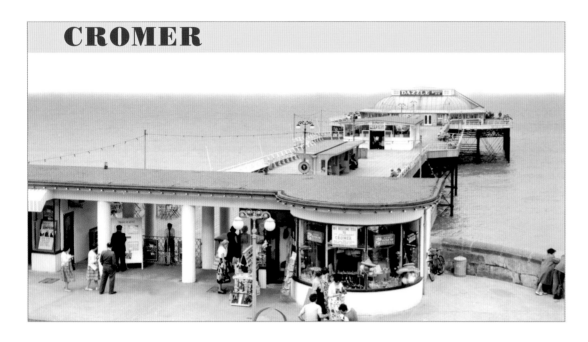

There had been earlier jetties here, but the seaside pier did not open until 1901. Built by the renowned contractor Alfred Thorne, it was a short and stubby pier compared with many others – just 500ft in length.

Constructed using substantial steel girders sitting astride strong cast iron columns, it was consequently low on maintenance costs, thus ensuring high profitability for the shareholders. A pavilion was functional in 1905, after the extension of a seaward end bandstand. It was later damaged by high seas in 1949 and 1953, and then restored and renovated. Traditional summer shows are still popular.

Fast stretching out to the east and west of the church that forms such a prominent landmark, Cromer stands high and bracing on its breezy cliffs, from which stairs and zig-zag paths lead down to the sands, which seem to be more and more crowded every summer.

It is fortified with an esplanade and breakwaters behind which it holds gallantly out. On undeveloped tastes Cromer would be thrown away. The cliffs are brown and sandy, the sea blue and the landscape of a universal green.

Cromer stands high and bracing on its breezy cliffs … on undeveloped tastes it would be thrown away…

Top left: CROMER, THE PIER C1960   C192046

Below left: CROMER, THE PIER 1902   49062

GREAT YARMOUTH, BRITANNIA PIER 1894   33385

GREAT YARMOUTH

The usual associations of [this] Norfolk watering-place are less with greatness than with bigness, boisterousness, and a joviality unrestrained by any false pride…

On page 33 is a rare view of the first Britannia Pier, which opened in 1858. It was dogged by disasters, including a ship collision – something piers were often vulnerable to – and storm damage. The structure was finally pulled down in 1899, though work on a replacement began the following year.

The new Britannia Pier (left, 52337) opened in 1901 with a temporary pavilion, which was pulled down to make way for a permanent pavilion a year later. It fell victim to fire in 1909, though it was replaced. This too was destroyed by a blaze in 1914, which was allegedly started by the Suffragettes, who had been refused permission to hold a meeting there. A third pavilion opened within months.

The view of the entrance (below, G56015) shows just how much of the Britannia Pier is really land-based. Today the Pier Tavern and amusements are sited here, with a pavilion theatre – the fourth one on the site after another fire – further out to sea. Unlike the Wellington Pier, Britannia has always been privately owned.

The first Wellington pier (below, G56036), built in 1853, was completely reconstructed after the council bought it. It re-opened in 1903, and included an impressive pavilion. The council also had the Winter Gardens transported from Torquay, and these became part of the pier complex, though they never went over the sea. Both the Winter Gardens and the Pier Theatre remain in use.

This is the most important town and port on the East Anglian coast; it is situated at the mouth of the Yare. Briefly, its attractions include firm and extensive sands for bathers, a marine parade, three piers, the Theatre Royal, and an aquarium. The older part of the town adjoins the river, and contains numerous picturesque 'Rows' or lanes, scarcely more than from 3 to 6 feet wide. There is a marine drive, three miles long; and the climax of the Yarmouth saturnalia is reached with the Regatta in August, when the town seems filled with what one may term the concentrated essence of Bank Holiday. Although it styles itself Great Yarmouth to distinguish it from that small Yarmouth in the Isle of Wight, the usual associations of the Norfolk watering-place are less with greatness than with bigness, boisterousness, and a joviality unrestrained by any false pride.

Left: GREAT YARMOUTH, BRITANNIA PIER 1904   52337

Above right: GREAT YARMOUTH, BRITANNIA PIER
C1955   G56015

Below right: GREAT YARMOUTH, WELLINGTON PIER
C1955   G56036

Like Great Yarmouth, Lowestoft has two piers. The older is the South Pier (right, L105079t), built in 1846 as part of the harbour. A reading-room was added in 1853–54, and a bandstand jetty in 1884. Both the bandstand and reading-room were destroyed by fire in 1885, though a new pavilion incorporating a reading room – shown in 37936 and 37337 on pages 38 and 39 – was constructed in 1889–91.

The reading-room was badly damaged during the Second World War, and its remains were demolished. A new pavilion was opened by the Duke of Edinburgh in 1956. However, this was pulled down in the 1980s as part of an ultimately unsuccessful marina project. What with the adjacent Claremont Pier having had its decking closed, Lowestoft's piers have seen better days.

Above: LOWESTOFT, THE SOUTH PIER C1955 L105100

Right: LOWESTOFT, THE SOUTH PIER AND PUNCH AND JUDY C1955 L105079t

LOWESTOFT

Opposite: LOWESTOFT, THE SOUTH PIER READING-ROOM 1896   37937

Above: LOWESTOFT, THE SOUTH PIER FROM THE SANDS 1896   37936

Below: LOWESTOFT, THE PIER MINIATURE RAILWAY C1955   L105076

In more ways than one Lowestoft and Yarmouth have long been rivals. Lowestoft is a more select resort; and there is no place on the coast of East Anglia that has more claims on the favour of strangers who seek an invigorating air, a pleasant neighbourhood, and an abundant means of amusement without too much noise and crowd. The harbour, almost entirely reconstructed, is protected by two piers, with a lighthouse showing a red light all night long at the end of each. 'I shall always look upon Lowestoft', says Mr Clement Scott, 'as the very pink of propriety. It is certainly the cleanest, neatest, and the most orderly seaside resort at which I have ever cast anchor. There is an air of respectability at the very railway station - no confusion, no touting, no harassing, and no fuss. I do not think I ever saw so neat a place out of Holland'.

SOUTHWOLD

Southwold Pier opened in the summer of 1900. Here the elegant 'Belle' Steamers drew alongside discharging and collecting visitors in what was the glorious era of Edwardian tourism.

Like all piers, Southwold underwent many adventures over the years but today it lives and breathes again as a real attraction for people of all ages and tastes, but especially for those with a sense of fun.

The photograph opposite (S168032) shows a quiet day at the pier; the fluttering flag perhaps indicates why. The pier had been reconstructed after wartime destruction, but in the year of the Frith photographer's visit an October storm caused severe damage, causing a section to be washed away, and reducing the overall length from 800ft to 370ft. The problem was compounded by neglect and further storms until 1999 when its rescue commenced.

Charms the visitor whose soul does not pine for a brass band and a general uproar

Southwold remains a quiet, sleepy, picturesque and wholly delightful retreat, innocent alike of piers and minstrels, of ugly terraces, and of those indications of 'life' which induce lovers of peace and the picturesque to fly in dismay from popular resorts by the sea. There is about Southwold a curiously unconventional air which instantly charms the visitor whose soul does not pine for a brass band and a general uproar. The whole town seems to overflow with sunshine, which lights up every corner of its rambling, unconventional streets.

Above left:
SOUTHWOLD,
THE BEACH 1919
69118P

Below left:
SOUTHWOLD,
THE PIER WITH THE
LONDON BOAT
LEAVING 1906   56833

Opposite above:
SOUTHWOLD,
THE PIER ENTRANCE
C1955   S168032

F elixstowe's pier opened in 1906 with a length of half a mile; the pier incorporated an electric tramway – one of its cars is clearly visible in the photograph. Alas, the tram service ceased with the advent of war in 1939, while the pier's seaward section was demolished after the war was over. Today's pier measures just 450ft.

Note the car and the group of children. They look as if they are there, but they have in fact been added to the photograph. Motor cars were the new invention and the Frith photographer was keen to make sure his photograph was right up to date.

FELIXSTOWE

Above right: FELIXSTOWE, FROM CONVALESCENT HILL 1904   51250
Right: FELIXSTOWE, THE PIER 1906   54640

With a southern aspect over the estuary of the Orwell, and enough shelter to give it a milder climate than most places on this Suffolk coast, Felixstowe has of late years risen rapidly, and already bids fair to outstrip its old rivals. The visit of the German Imperial Family certainly did much for the place by proclaiming the merits of its sea-bathing. The single line of the Great Eastern railway to the town was opened in 1877. The summer excursion fares are amazingly low; the courteous officials and fine stations a standing reproach to certain other great railway companies we could mention.

WALTON-ON-THE-NAZE

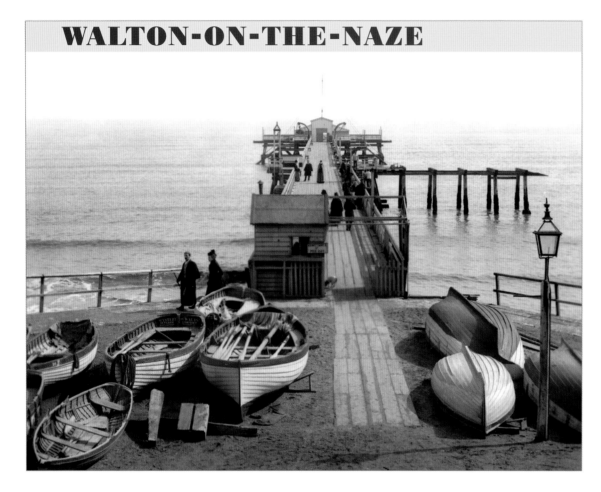

Walton-on-the Naze is set alongside the breezy North Sea, a little along the coast from its larger and more worldly neighbour Clacton. It has had two piers: the first, shown in the photograph, was built in 1830 as a simple wooden jetty for the landing of passengers and cargoes from steamers. It suffered severe storm damage and was finally destroyed in the early 1890s.

Walton was growing rapidly in popularity and, in 1895, just a few years after the loss of its pier, a new 800ft pier was constructed to the designs of J Cochrane by the Walton-on-the-Naze Hotel and Pier Company. This stronger and more sturdy structure was added to considerably down the years, and Walton now boasts the third longest pier in Britain, its overall length 2,600 ft.

A single-track tramway was installed in 1898 to carry passengers back and forward from the boats, and operated up until the 1930s. This was replaced first with a battery-powered carriage, and later by a diesel locomotive. Today, unfortunately, there is no pier railway.

Left: WALTON-ON-THE-NAZE, THE PIER 1891 28237

Now largely resorted to by Londoners

This small but rapidly extending watering-place, washed on two sides by the sea, was formerly frequented in autumn by the gentry of Essex and Suffolk. But it is now largely resorted to by Londoners, who have virtually to thank the Great Eastern Railway Company for 'discovering' the place, and handing it over to the tired millions of our Metropolis. There is at Walton a smooth, sandy beach, several miles in extent, which is a veritable paradise for bathers and children. From the Crescent Pier steamboats ply to London and Ipswich.

SOUTHEND-ON-SEA

HMS 'Leigh'

At the beginning of the Second World War the Royal Navy requisitioned Southend's pier and renamed it HMS 'Leigh'.

They installed pill boxes and anti-aircraft guns, and were equipped to drop depth charges on to maurauding enemy submarines. However, there was only a single attack of any significance – a strafing by the Luftwaffe. Convoys embarked from the pierhead and the pier railway was used to transport injured and wounded soldiers from ships to hospitals inland.

SOUTHEND-ON-SEA, THE PIER 1898   41377

This popular resort can be heartily commended to all, but especially to Londoners. It is reached in little more than an hour by the excellent trains of the Great Eastern Railway. It is quite remarkable to see the crowds of Londoners poured into Southend by steamboat and excursion train on a fine summer's day. The coast here is very shallow, and the tide retires nearly a mile from the shore at low water. The old town stretches along the shore eastwards from the pier in a line of shops and small houses inhabited by the boatmen and fishermen, who make up the mass of the population.

SOUTHEND-ON-SEA, THE PIER C1962   S155102

The longest seaside pier in the world, Southend's first pier lasted from 1830 to 1887. It was then replaced, and the new structure opened on 24 August 1890. It cost £42,000. Extensions were opened eight years later which took its length to a record-breaking 7,080ft. An electric railway took people to the pierhead.

Southend Pier had become so popular that the rail track was doubled in 1929, and the Prince George steamer extension was built. During the Second World War it became HMS 'Leigh'. Afterwards a new electric train catered for millions of visitors, many from the East End of London, for whom Southend was their special resort.

By 1960, visitor numbers had halved from the almost 6 million of the pier's post-war peak. 20 years later, local councillors planned closure, but a last-minute rescue ensured its future. A new pier railway was opened in 1986 by Princess Anne. Southend was the favourite pier of Sir John Betjeman, first President of the National Piers Society.

Clacton's wooden pier opened at the height of the pier boom in 1871. It had been built by Peter Schuyler Bruff, the manager and engineer of the Eastern Union Railway. Soon trains were arriving from Liverpool Street in London, and paddle steamers including the 'London Belle' and the 'Clacton Belle' were wishing to berth at the pier to disgorge further visitors. Extensions were necessary, and they were created between 1890 and 1893, and masterminded by Kinipple and Jaffrey. New buildings included a new polygonal head and landing stage, a camera obscura, and a pavilion for entertainments. The length was thereby increased to 1,180ft. Steamer services were stepped up and a timetable on the Ladies' Baths advertised services to Felixstowe, Harwich, Ipswich, Yarmouth, Southend and Gravesend.

In 1885, 94,000 pier tickets were sold, and just four years later the number had shot up to 147,000.

It was not until Ernest Kingsman bought the pier in 1922 that the series of inter-war amusement buildings associated with Clacton pier were added. The pier is much altered in recent years, but it is still amusement-dominated.

Right: CLACTON-ON-SEA, THE PIER C1960 C107051

Below: CLACTON-ON-SEA, THE PIER C1960 C107062

The town has a southern aspect, and it commands a fine view of the German ocean. It stands on cliffs some 40 feet high, directly facing the sea. For cleanliness, firmness, and extent, the sands of Clacton-on-Sea cannot be excelled; they are a perennial source of delight to the children, and for visitors generally they form a safe and pleasant promenade for miles each way. Bathing is safe at all times, the sands being firm, and sloping gently out to deep water.

CLACTON-ON-SEA, THE BANDSTAND AND THE PIER 1907 58934

Herne Bay's first pier was built by the engineer Thomas Telford in 1832. With a distinctive T-shaped head, it was almost three-quarters of a mile in length (3,640ft). Its extreme length was made necessary by the shallow waters along the coast here, which made it impossible for steamers to venture in too close to the shore. Fortunately, passengers were able to transfer their heavy luggage on to a sail-powered trolleyway. Steamer services declined during the 1860s, causing the pier to deteriorate, and it was finally demolished in 1870. A second much shorter pier was built in 1873, with a theatre and shops for the benefit of visitors. This structure was added to and extended in the late 1890s, until finally Herne Bay could boast the second longest pier in the country, after Southend.

In February 1953 the resort's promenade was buried under thousands of tons of shingle, and the roads blocked by smashed boats. It was caused by the worst floods the county had known when a tidal surge broke through the sea defences. The pier, badly damaged, never recovered, and the neck finally succumbed to bad weather in 1978 and broke apart, leaving the head isolated far out in the water.

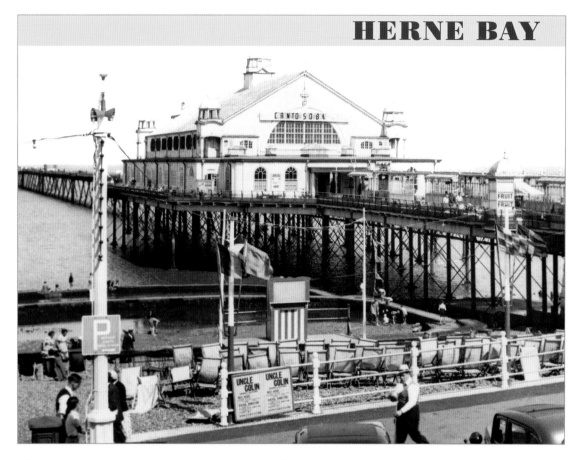

HERNE BAY

Above: HERNE BAY, THE PIER 1953 H75025   *Below:* HERNE BAY, THE PUNCH AND JUDY SHOW C1955 H75019

This town dates its history as a watering-place from the year 1830. Extensive speculation in building was followed by failure; but the place has entered on a new lease of life since the advent of the London, Chatham and Dover Railway in 1862. Herne Bay commands a magnificent view over the North Sea. The fine clock tower was the gift of a private lady, and cost £4,000. The sea air at Herne Bay is considered to be more bracing than at any other resort in England. Not very long ago people said Herne Bay was too quiet – even dull; we are quite sure the reproach is unjustifiable now, for there has been a great awakening in the town.

Though called the Jetty, the structure shown in 68444 (right) was a fully-fledged seaside pier, unlike the harbour wall that locals knew as the pier. Margate has long been a seaside resort, with claims that it had a landing jetty as far back as 1800. The structure illustrated was the first iron pier, which opened in 1855.

The pier was also renowned for being the first to be designed by Eugenius Birch. Extensions were made in 1875–78, when the octagonal pierhead and pavilion were added. Further developments were made in 1893 and 1900. In 1897 the pier company recorded a profit of £1,689 from visitors and landing charges levied on steamers and yachts.

Like Brighton Chain Pier, Margate Jetty once had a camera obscura. Alas, the jetty was virtually destroyed by a storm on 11 January 1978, after having closed two years earlier on safety grounds. Part of the isolated pierhead still survives as a rusting tangle.

Blame the toll collector

In 1812, a toll was levied on visitors to the landing jetty, which resulted in riots. The toll collector was threatened by the angry hordes and was almost hurled into the sea.

Walking out to the end of the jetty in the early 1800s could be hazardous. At high water it was totally covered by waves. Unsuspecting promenaders were often cut off and left stranded by the rising tide, and had to be rescued by fishermen, who charged a fortune for their services.

Margate's cliffs are bold and picturesque. Yet 'superior' people think it needful to offer some excuse for their being found at such a place, and are at pains to explain how they must by no means be confounded with the ordinary ruck of Margate's guests. There can be no question as to the past respectability of Margate. The narrow High Street, the old houses hidden away here and there, the shallow harbour – all these show that the town was once a fishing port. There is no affectation, no blasé cynicism about your genuine Margate visitors. Its sands are thronged by a crowd of idlers ready to be easily entertained by jugglers, Punch and Judy shows, and wandering minstrels. There are busy vendors of refreshments and knick-knacks; family parties, encamped with umbrellas and novels; eager children, sprawling babies and their nurses, and scores of adventurous youngsters, wading in the surf.

MARGATE

Right: MARGATE, THE JETTY 1918   68444

PEGWELL BAY

This little-known structure was built in 1878 for a local reclamation company, replacing an earlier jetty. Its main purpose was intended to be a landing stage for steamers, but few ever called. The company ended up going bankrupt, and the pier was pulled down inside a few years.

Pegwell Bay is celebrated for being the point of landing for Jutish invaders led by Hengist and Horsa. Today this tiny resort boasts a replica Viking dragon-headed longship, the 'Hugin', which was rowed and sailed from Denmark to Broadstairs by 53 Danes to celebrate the 1,500th anniversary of the invasion. It appeared on the cover of the Stranglers' 1979 album 'The Raven'. Here, too, the Pre-Raphaelite William Dyce painted his famous painting of the bay in 1858, showing the first tourists gazing into rock pools and enjoying the sunlight.

Above: PEGWELL, THE SEAFRONT AND THE PIER C1880   12739
Left: PEGWELL, THE DANISH VIKING SHIP, HUGIN C1960 P20021

All in the line of duty

At nearby Ramsgate, a detective from London was enjoying a break from police routine by promenading in the sun on the town's pier. He happened to look into the camera obscura, only to spot a well-known local villain working his way through the crowd picking pockets. Dashing back to the shore he rapidly apprehended him.

DEAL

In all weathers strangers should look out for the place where the sewage is discharged …

To many people, the principal attraction in this pleasant old Cinque Port will be the ever-changing view of shipping in the Downs. Behind the shelter of the Goodwins, the tourist may enjoy safe sailing by the famous boatmen of Deal. A particularly steep and shingly beach, some three miles in length, affords a capital bathing-ground, duly provided with machines and tents; but caution is necessary in rough weather. In all weathers, however, strangers should look out for the place where the sewage is discharged. There is a stretch of sand at one end where children may disport themselves in safety. The Deal boatmen, limited by statute to the number of fifty-six, are famous for their heroism. Their skill and daring are often put to the proof on this dangerous coast, for beyond the sheltered Channel near the Downs – a vast natural harbour, eight miles long and six miles wide – lie the treacherous Goodwin Sands, from three to seven miles broad, and at low water these are so hard and firm that cricket matches may be played upon them.

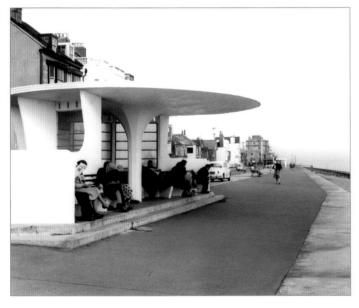

This Kentish resort has had three piers over the years, although not concurrently. The first one, designed by Rennie, opened in 1838, but it decayed owing to storm damage and sandworm attack. It was washed away, and was replaced by a Birch-designed pier that opened in 1864. A pierhead pavilion followed in 1886.

Birch's pier was a victim of the Second World War: only the tollhouse remained after the damaged structure had been pulled down to assist the needs of a coastal gun battery. However, it was replaced by a reinforced concrete structure, built from 1954 to 1957. It remains council-controlled, and popular with anglers.

Opposite: DEAL, THE PIER C1955 D15087

Top: DEAL, THE BANDSTAND AND THE PIER 1906 56911

Above: DEAL, THE PROMENADE C1960 D15092

The most celebrated of Britain's pier designers was Eugenius Birch (1814–1884). Early in his career as an engineer he was successfully building railways, viaducts and bridges, as well as furthering the cause of the Empire by working on the Calcutta–Delhi railway line in India. He made two vital contributions to the design of piers: the first was his introduction of the exotic styles of decorative architecture he saw in the Indian sub-continent. Suddenly piers were light-hearted, exciting and beautiful. The second, the use of screw-piles, was very different: up to then, pier supports had been simply wooden posts hammered into the seabed. Birch cunningly attached screw blades to the bottom of iron piles and screwed them into the bed, achieving tremendous strength.

DOVER

The Admiralty Pier (48059, left) was a traditional steamer pier; cross-Channel ferries began operations in 1851, nine years before the railway's arrival. The pier was subsequently widened in the 1910s after Dover Marine station opened, initially just for military traffic. In 1994 the rail service ceased, though ships still go from here.

The pier shown in 48060 (below) was called the Promenade Pier to distinguish it from the Admiralty and Prince of Wales piers, which were both primarily used by steamers and trains. The Promenade Pier opened in 1893, and was 900ft in length. A ship collided with the structure in the same year and caused substantial damage, and heavy seas and storms during the next winter added to its delapidation. However, a pavilion was added in 1899 and Dover's pier regained its popularity with visitors for a further quarter of a century. However, by 1925 it had become dilapidated once again. Repair costs were felt to be unjustifiable, and demolition followed in 1927.

One of the Cinque Ports, Dover is about two hours run from London by express train. During the summer months there is a good service of steamboats between this interesting watering-place and London. The harbour may be divided into three parts, namely: the pent or breakwater, the basin, and the outer harbour. Favourite walks are on the heights to the castle and to Shakespeare's Cliff, which commands a broad view of the shores of France. The castle, as it stands, is practically of the date of Henry II.

Left: DOVER, ADMIRALTY PIER 1901 48059

Above: DOVER, THE PROMENADE PIER 1901 48060

FOLKESTONE

Folkestone's Victoria Pier opened in 1888, and included a 700-seater pavilion. Famous names that performed here included Lily Langtry, associated with the future Edward VII, and the clog dancer Dan Leno. This high class variety proved expensive, so new leasees introduced less costly entertainment such as beauty contests and film shows.

During the First World War, the pier was popular with troops. However, it remained closed for the entire duration of the Second World War. On Whit Sunday 1943, a fire wrecked the pavilion; the rest of the pier was also badly damaged. Its ruins were demolished in 1954.

The policy of the townspeople is such as to discourage excursionists and seek the patronage of the higher class of visitors ...

The older portions of the town have steep and narrow streets, but the modern houses on the cliffs are most attractively situated, fronting the well-known promenade called the Leas, from which one may easily reach the beach. The cliff pathway, between this part of Folkestone and Sandgate, is one of the most beautiful walks in the kingdom for lovers of coast scenery. The season here is short and late, and the policy of the townspeople is such as to discourage excursionists and seek the patronage of the higher class of visitors; consequently, the humours of the sands so conspicuous at Margate or at Yarmouth, are hardly to be looked for at Folkestone.

Folkestone pier is said to have been the venue for the first ever international beauty show in Britain. Local girls preened and pranced for the delectation of visitors, and the organisers shrewdly spiced up the affair by bringing a number of attractive French girls over from Boulogne, much to the delight of the local male population.

Opposite: FOLKESTONE, THE LEAS FROM THE PIER 1901 48054

Top: FOLKESTONE, THE VICTORIA PIER 1895 35530

Above: FOLKESTONE FROM THE AIR 1931 AF35417BL

HYTHE

This was always more of a landing pier; it opened in 1881 for use by the Hythe to Southampton ferry. Though Hythe Sailing Club used to have a clubhouse here, its main facility has long been the railway, which developed from a goods-only baggage line installed in 1909.

The 'Hotspur II' is seen here (H372084, above) connecting with the train, currently operating a half-hourly service. Converted to take passenger traffic, it has used the same locomotives ever since the line opened in 1922. Extensive replanking of the pierhead was carried out in 1982, after development to the buildings there in 1970–71.

Above: HYTHE, THE FERRY C1960 H372084

Left: HYTHE, THE PIER TRAIN C1955 H372025

Right: HASTINGS, FROM THE AIR 1931 AF35334

HASTINGS

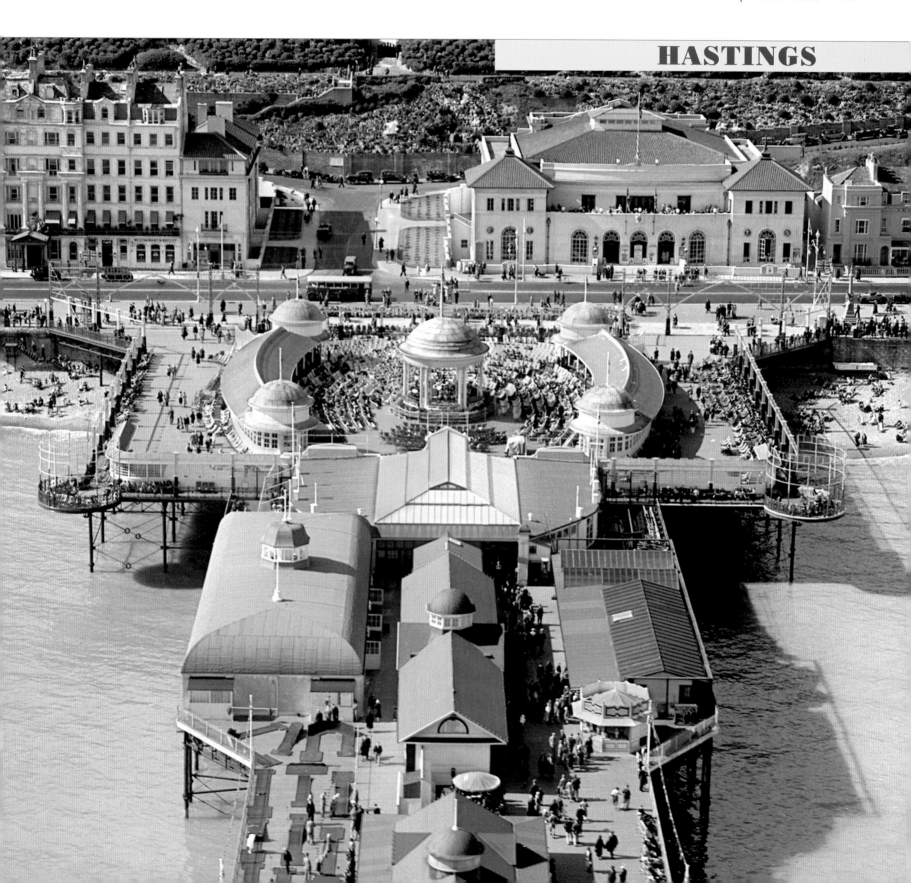

This is another pier by Eugenius Birch. Costing £23,000, it opened in 1872, and included a seaward end pavilion capable of housing 2,000 people. It was originally some 900ft long – a dramatic structure for strolling along and for seaside entertainments. The fine pavilion at the end of the pier was destroyed by fire in 1917. A landing stage was added in 1885. A notice outside one of the tollhouses, damaged by a storm in 1877 but repaired, shows that it would have cost you 2d to walk to the pierhead.

Further development at the shoreward end included a rifle range and bowling alley. Known as the 'parade extension', this was sold to the council in 1913 to finance a new arcade, shops and a tea-room. The pavilion here is a 1922 replacement which was built after the original structure had been destroyed by fire. New owners took over in 2000.

Hastings lies – for the most part – in a hollow, snugly sheltered by hills, except where it slopes southward to the sea. Of course, the increase of houses of visitors must tend to spoil the natural freshness and original individuality of a population, but in Hastings these qualities are preserved to an unusual extent, especially among the fishermen. Under the East Cliff, 'Dutch' fish auctions are often held. The bathing here is excellent in every way. Under the Parade near the pier are fine baths, erected on the foreshore in 1879 at a cost of £60,000. Lord Byron and Charles Lamb have left records of their stay here, the latter describing his sojourn as 'a dreary penance'.

Page 61:
HASTINGS, FROM THE
AIR 1931 AF35334

Opposite:
HASTINGS, THE VIEW
FROM THE PIER C1955
H36052P

Above:
HASTINGS, THE PIER
1925 77984

Right:
HASTINGS, THE PIER
1890 22780

ST LEONARDS

Built at a cost of £30,000 in 1891, St Leonards pier lasted for just 60 years. Its main feature was the sumptuous 600- to 700-seater pavilion shown in the photograph, designed by F H Humphreys.

This historic view shows the 950ft-long pier in the year that it opened, with its short-lived landing stage.

During the Second World War the structure was damaged by enemy action, and was finally demolished in 1951.

Announcement
ST LEONARDS PIER
Saturday 3pm
**EXHIBITION OF
BEAUTIFUL CHILDREN**
*Handsome prizes for those
adjudged the most beautiful.
Children competitors free.
Visitors 2d.*

Above: ST LEONARDS, THE PIER
1891 29607

EASTBOURNE

EASTBOURNE, THE PIER 1925   77946T

There were fears during the Second World War that enemy forces might take advantage of piers and use them as convenient landing stages to mount an invasion. During one performance, members of the audience in the pier theatre were distracted by sounds of banging and clanging from outside. The performance came to an abrupt halt. Everyone was astonished and more than a little alarmed to discover teams of soldiers energetically fixing explosive charges to the structure.

Designed by Eugenius Birch, the pier opened in 1870. Its first theatre seated 400, and cost a mere £250 – it eventually became a cattle-shed at Lewes! The saloons visible here halfway along the decking were added in 1901, the same year that work on the new pavilion was completed.

Would an artist billed as 'The Gay Lord Quex' perform today (64968, below)? The year 1925 marked the building of a new 900-seater music pavilion at the shoreward end of the pier. As can be seen in photograph 77946t on page 65, already coach (or charabanc) traffic was having an impact, bringing visitors to sample the pier's delights. We may be thankful that although a later entrance building was wrecked by fire, both the older theatre and music pavilion seen in this photograph can still be seen today, and the pier remains successful.

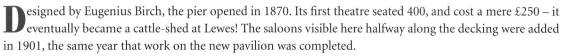

One must mention that there are noble tree-planted streets and shady avenues, an imposing sea-front of about three miles, an excellent beach of mingled sand and shingle, a pier of the most approved pattern, gardens and promenades, and every convenience for bathing, boating, and fishing, as well as first-class hotels, well-built houses, tempting shops, and irreproachable sanitary arrangements and water supply.

Left: EASTBOURNE, THE PIER 1910   62958P   *Above:* EASTBOURNE, THE PIER 1912   64968 (DETAIL)

The well-known early Chain Pier at Brighton was engineered by Captain Samuel Brown, who had designed a smaller but similar structure, which opened at Leith near Edinburgh in 1821. The Chain Pier lasted from 1823 to 1896, gradually falling victim to storms, neglect and a loss of business to the West Pier.

By the 1890s it was in a sorry state, having been beaten by the Channel storms and high winds for 70 years. When surveyed it was discovered that the structure had slumped over six foot out of the vertical. Now potentially dangerous, it had to be closed. In 1896 it was again ravaged by storms and collapsed entirely.

Left: BRIGHTON, THE CHAIN PIER 1870 B208003

Below: BRIGHTON, THE WEST PIER 1896 33717

The fascinating flea circus

Brighton's West Pier was the height of fashion in the late 19th century, and certainly the place to be seen. However, it was famous, too, for its unusual amusements and sideshows. In 1890 lucky visitors were treated to a flea circus. Fascinated onlookers enjoyed the sight of 'a morning drive with carriage drawn by fleas', and a cannon that was set off by a flea.

BRIGHTON

Was Brighton's West Pier the finest pier of them all? It was deemed worthy of Grade One listed status by the Department of the Environment, indicating its national importance. It opened in 1866, with the seaward end pavilion dating from 1893.

Like many piers, there always seemed to be new developments taking place here. The southern pavilion was rebuilt and extended only two years after its initial construction. Landing stages were built in 1896, and these were also extended in 1901.

The penultimate stage of building was in 1916 – strangely enough, during the First World War – when a bandstand midway down the pier was removed as part of a widening process. A replacement Concert Hall followed. Only a new top-deck entrance remained to be added.

Above: BRIGHTON, FROM THE WEST PIER 1902 48497

Right: BRIGHTON, FROM THE WEST PIER 1921 71486

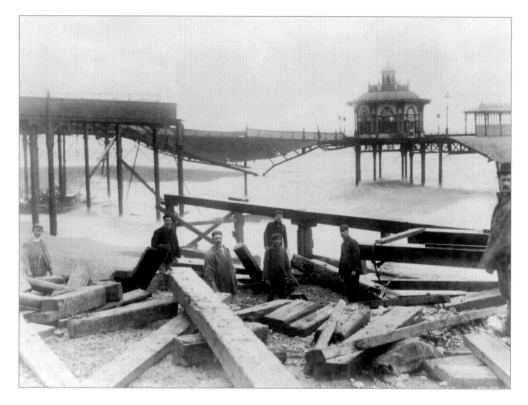

The 1896 storm that destroyed the Chain Pier also badly damaged Brighton's partly-built Palace Pier and the West Pier, shown here (B2085009). Was this a foretaste of what was to come? Sadly, it seems so. Despite being taken over by a trust and gaining lottery funding for its restoration, objections, storms and arson damage appear to have sealed the fate of Eugenius Birch's masterpiece.

Pier Disaster

In stormy weather in October 1973 a barge, moored to the pierhead of West Pier, broke free. It repeatedly crashed against the western side of the structure, causing damage to a number of the supporting piles. Horrified onlookers watched the helter skelter tottering, until it, the crazy maze and much of the floor decking tumbled down into the water. The pier theatre did not escape the tragedy, one of its corners hanging unsupported over the edge. The English, however, always relish a disaster, and visitors were soon paying to crowd through the turnstiles to inspect the damage.

Though it opened in 1899, the 1,800-seater theatre at the southern end of the Palace Pier was not completed until 1901 – a full decade after work on the pier began. A fine pavilion and winter garden were added to the pier's centre in 1910. The theatre was dismantled in 1986, but was not replaced as had been planned.

Opposite above:
BRIGHTON, THE WEST PIER
C1896 B2085009

Opposite below:
BRIGHTON, THE PALACE PIER
1902 48513

Left:
BRIGHTON, THE PALACE PIER
C1955 B208022

Can anyone now realize a Brighton with only 1800 inhabitants, and those mostly poor fishermen? Yet that was the Brighton of little more than a century ago. This watering-place owes its present prosperity, in the first place, to a physician, Dr Russell, of Lewes, who removed hence in 1750. He published a treatise on the advantages of sea-bathing, recommending Brighton very strongly. The Aquarium is situated between the Steyne and the Chain Pier, and was erected by a joint stock company at a cost of £130,000 in 1872. We rather think it is more of a promenade than an aquarium, with its elegant corridors, conservatory, and saloons provided with newspapers, periodicals, and the latest telegrams. There are forty-one fish tanks. You can scarcely move on the Parade on a fine afternoon without meeting troops of fair horse-women attended by their riding-masters, sweeping along, perhaps, towards the Downs. The stream of carriages is almost as incessant as on a Drawing Room day at Buckingham Palace. Bands are playing wherever you go, till the very air grows musical.

WORTHING, THE BANDSTAND AND THE PIER 1921 71445

That Worthing has a milder climate than its neighbours is shown by the large quantities of fruit and vegetables which it sends to Covent Garden. Besides being a quiet holiday resort in summer, Worthing is well adapted for delicate persons in winter, when the flourishing lauristinus hedges still brighten its streets. Complaint, however, is made of fogs, and still more strongly of the seaweed, which accumulates on the beach here in such quantities as to become a perfect nuisance. Everyone is aware of the misfortune which befell Worthing, when its water became tainted, and the demon of typhoid descended upon the town. That cloud, however, has now passed away, and the drainage is above reproach. The sands are smooth and hard, and their condition during the summer months has been graphically described as one long mile of nursery. There are many people who prefer this watering-place to Brighton, on the grounds that it is quieter and far more economical to live in.

WORTHING

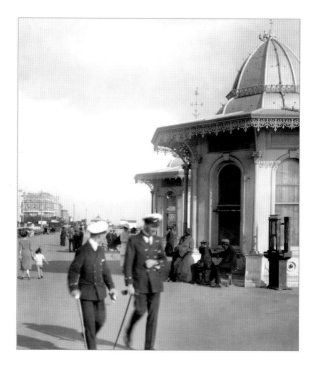

There has only ever been the one pier at Worthing. The structure, built in 1862, is currently the sixth oldest surviving pier. Its first entertainments were provided by a 9-piece band, which performed here in 1874. The decking was widened in 1888–89, when a 650-seater pavilion was added at the pierhead.

Steamers used to call at the landing stage, but they were hampered by the low tides. Plans to double the pier's length to take more boat traffic were abandoned when it was realised that this would only make the water 6 inches deeper. Easter Monday 1913 saw the pavilion left isolated when the decking collapsed - London's Lord Mayor carried out the re-opening a year later.

Two kiosks had been added to the pier's entrance back in 1900, and these are shown in 71457 (left). The pier was taken over by Worthing Corporation in 1920, with a new shoreward end pavilion opening six years later. The pier's southern pavilion became a fire victim in 1933, but was replaced. A further amusement pavilion opened in 1937.

Troops were billeted at the shore-end pavilion during the Second World War, though it had re-opened to the general public by 1946. A facelift was carried out to this pavilion in 1958, with a £1.1 million renovation scheme taking place in 1979-82. Today the Pavilion Theatre houses dances and concerts, both of the classical or rock music variety.

Top: WORTHING, THE PIER HEAD 1921 71457 (DETAIL)　　　*Above:* WORTHING, ON THE PIER C1955 W147048

BOGNOR REGIS

The pier opened in 1865, its entire length 1,000ft. Costing £5,000, it was intended for simple promenading, for there were few buildings. A seaward end pavilion was constructed in 1900, and a new shoreward end complex added a decade later, including a cinema, arcade and theatre.

By 1911 (left and below) the pier had acquired an agglomeration of buildings at the landward end to replace the elegant 1865 entrance kiosk with its ogee pyramid roof; its width had been increased to 80ft. They were built by the Bognor Pier Company, who purchased the pier from the council in 1909; the complex contained a cinema, theatre and arcades of shops.

Today the seaward end pavilion is no more, a victim of storms in March 1965, when it broke up and fell into the sea. This threatened pier, which was unfortunately refused Lottery funding, hosts an annual Birdman Rally.

This town is Worthing's twin sister – a quiet, mild, healthy watering-place, situate on a level in the face of the ever-restless Channel. About 1785, Sir Richard Hotham, a wealthy Southwark hatter, who determined upon acquiring the glory of a seaside Romulus, set to work to erect a town of first-class villas in this pleasant spot, with a view to creating a truly recherché watering-place, to be known to posterity as Hothamton. He spent £60,000; he erected and furnished some really commodious villas, but did not succeed in giving his name to his own creation, and died broken-hearted in 1799. For a Sussex watering-place Bognor is remarkably quiet, but it will doubtless commend itself to some people on this account.

Left: BOGNOR REGIS, THE PARADE AND THE PIER 1911   63793

Above: BOGNOR REGIS, THE PIER 1911   63786

SOUTHSEA

Above: SOUTHSEA, THE SOUTH PARADE PIER C1935   S161007

The South Parade pier is the more traditional of Southsea's two piers.

It was officially opened in 1879 by Princess Saxe-Weimar; its entire length was 600ft, and it was extensively rebuilt during the Edwardian era after a fire in 1904.

A spacious pavilion included a 1,200-seater theatre, which has since been replaced by a smaller structure.

The rock opera 'Tommy' was filmed here.

During the filming of the rock opera 'Tommy', a fire broke out in the wardrobe room. Rapidly, the entire cast and film crew evacuated the pier. Over 100 firemen battled with the flames. Although considerable damage had been done, they managed to save the pier, and rebuilding work began the next year.

Continually alive with yachts, steamboats, and battleships, it can never be dull ...

As practically the west-end of Portsmouth, Southsea holds a unique position among watering-places. It would not be rustic or romantic enough for all tastes, but recommends itself to many by the stir of military and naval life. What with regimental bands, parades, and reviews by land, and the Solent continually alive with yachts, steamboats, and battleships, it can never be dull; nor is it surprising that not a few old officers think there is no place like Southsea for a permanent or temporary residence.

Leading from the Esplanade to the shingly beach is the bathing stage of the Portsmouth Swimming Club, which is well appreciated by swimmers. One may reach the Isle of Wight by steamer in less than half an hour, and there are also excursions by water to Southampton and Bournemouth.

COWES

This is one of the shortest piers ever built at just 170ft; it opened in 1902. A pavilion was constructed two years later. Ferries ran to Southampton and Portsmouth, and the pier was used for the movement of troops during the First World War. The Royal Navy took it over for the Second World War.

The Victoria Pier (74750, above) was a hub of waterfront activity, with sailing clubs making full use of its facilities, especially during Regatta week. Note the attached banner – advertising of this nature was popular at the time. Ironically, this trend has not really been adopted in the more commercialised society of Britain as she is now.

Though the Victoria Pier is called the Old Pier here (C173011, right), Cowes' other seaside pier – Cowes Royal – had long gone. In fact, it only lasted fifteen years, from 1867 to 1882. Unfortunately, Cowes Victoria Pier was to follow suit. No funds were available for repairs, hence the 'no admittance' notice half-hidden by a kiosk. The last remnants of the structure were removed in 1965.

Top: COWES, VICTORIA PIER 1923 74750
Above: COWES, THE OLD VICTORIA PIER C1950 C173011

RYDE

RYDE, THE PIER C1883 16297

RYDE, THE PIER 1892   30033

This is the oldest true pier – it opened in 1814. Eventually 2,305ft long, there were at one stage three strands of pier running alongside each other, incorporating a railway and tramway, as well as a more traditional walkway. Ships landed at the head, where there used to be a pavilion (demolished in 1971) and refreshment room.

The tramway closed in 1964, but the railway remained. Passengers can travel from the pierhead through to Shanklin, travelling on old electric stock that used to run on the London Underground. A regular catamaran service goes to Portsmouth Harbour, with a café recently opened at the ferry terminal.

To the timid, and to those who are deterred by other causes than fear from venturing on the 'heaving wave', the pier affords innumerable attractions. The arrival and departure of the steam packets; the numerous boats everywhere sailing about; the merchantmen constantly underway; together with the occasional naval salutes, announcing the arrival or departure of ships of war, compose a scene of unusual interest and excitement.

Nor is the spectacle on the pier itself the least attractive object, from the number and often from the elegance and beauty of the fair promenaders. A more delightful scene can scarcely be conceived than this pier affords when the placid brightness of a summer's moon rests upon it. The combination of motion and stillness which the sea presents on a fine and tranquil night is inexpressibly pleasing.

SHANKLIN

The Isle of Wight has been amongst the unluckiest areas for piers, with only four of its original structures now standing. Shanklin, 1,200ft long, built for the Shanklin Pier Company, endured from 1891 to 1993; it would probably have been in existence today, had it not suffered severe damage from the hurricane which devastated southern England on 16 October 1987. The pavilion had already been destroyed in 1927, by a fire outbreak.

One cannot fail to notice the excellent taste displayed in the construction of villas and houses. People have complained that Shanklin does not 'go ahead' fast enough, but no one who loves it desires anything of the kind. Crowded assembly-rooms and inferior theatres are indeed poor substitutes for a moonlight view from Shanklin Downs, a view of the peaceful bay with its sea breaking in a low, hushed voice.

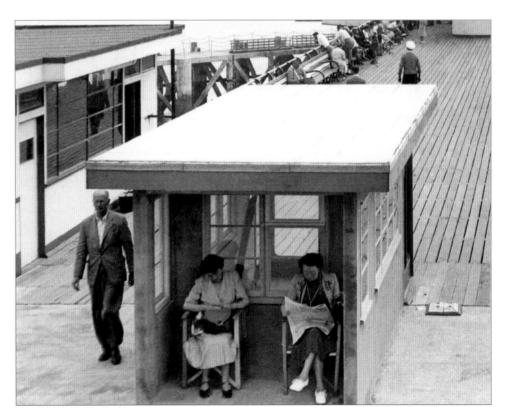

Left: SHANKLIN, THE BEACH AND THE PIER C1950 S104035
Above: SHANKLIN, THE PIER C1955 S104062

SANDOWN

The visitor to Sandown must not be repelled by the first aspect, which is somewhat bleak ...

The visitor to the Isle of Wight a few years ago would have found no Sandown at all, only a lovely horse-shoe bay paved with solid, shining sand, a few fishermen's cottages, and a half-drowned fort. Now it is one of the most popular resorts in the island, thanks to the railway, its lovely situation, and perfect bathing accommodation. Occupying a break in the line of cliffs of iron-sand and dark coloured clays, which form the sides of the ample bay, it enjoys the benefit of the inland breeze as well as the sea air, and is consequently less oppressively hot than Shanklin. The visitor to Sandown must not be repelled by the first aspect, which is somewhat bleak, owing to the absence of trees; and, moreover, as the cliff does not rise immediately behind the town, it can never be so pretty or so picturesque a place as Bonchurch or Ventnor.

It took a while for Sandown Pier to be erected: a Bill was passed by Parliament in 1864, but work did not begin until 1876. The pier finally opened three years later, measuring 360ft. Extensions in 1895 increased its length to 875ft, with a pierhead pavilion among the new attractions.

A landing stage opened the same year as the pierhead pavilion, catering for a generation of paddle steamers. Sandown Council took over the pier in 1918, and later amalgamated with their Shanklin neighbours. The new authority constructed a 1,000-seater shoreward end pavilion, which opened in 1934.

The pierhead pavilion remained in use as a ballroom, before eventually becoming a victim of fire. Twinkle-toed holidaymakers were able to 'Dance to the Melotones'. The pier survived a £2 million blaze on August Bank Holiday Monday 1989, and is a commercial success. However, the theatre inside the redeveloped shoreward end pavilion ceased operating in 1997.

Opposite: SANDOWN, THE PIER C1955   S57022   *Above:* SANDOWN, THE PIER 1935   86901

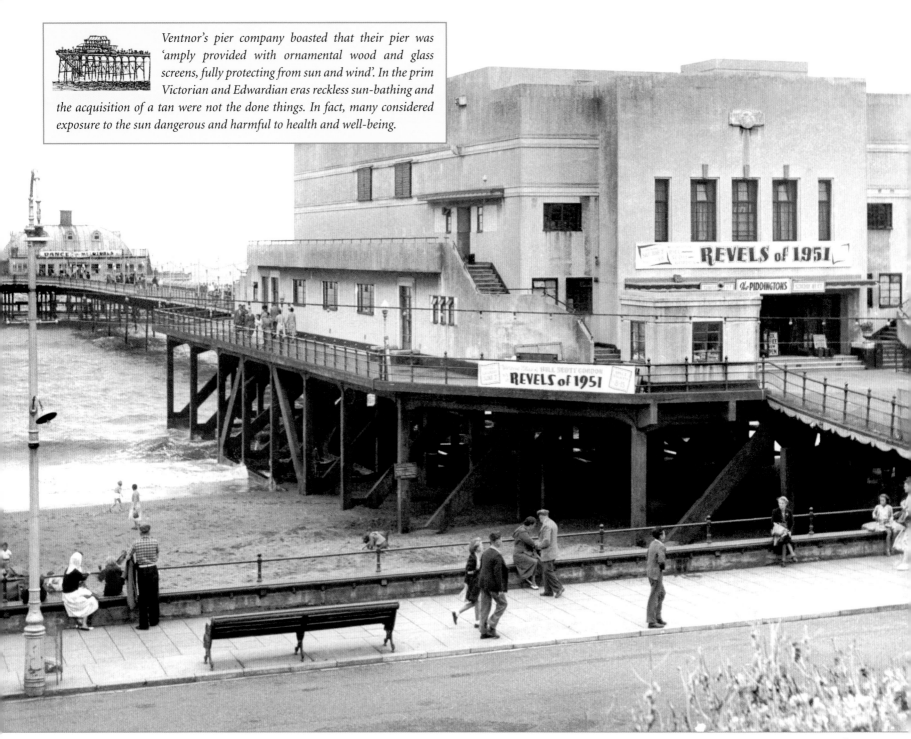

Ventnor's pier company boasted that their pier was 'amply provided with ornamental wood and glass screens, fully protecting from sun and wind'. In the prim Victorian and Edwardian eras reckless sun-bathing and the acquisition of a tan were not the done things. In fact, many considered exposure to the sun dangerous and harmful to health and well-being.

SANDOWN, THE PIER C1951   S57020

Though the landing stage decayed during the Second World War, and the pier was breached for defence reasons, Sandown pier soon re-opened for business afterwards. Indeed, in the year of the Festival of Britain, it had its own 'Revels of 1951' to pull in the visitors.

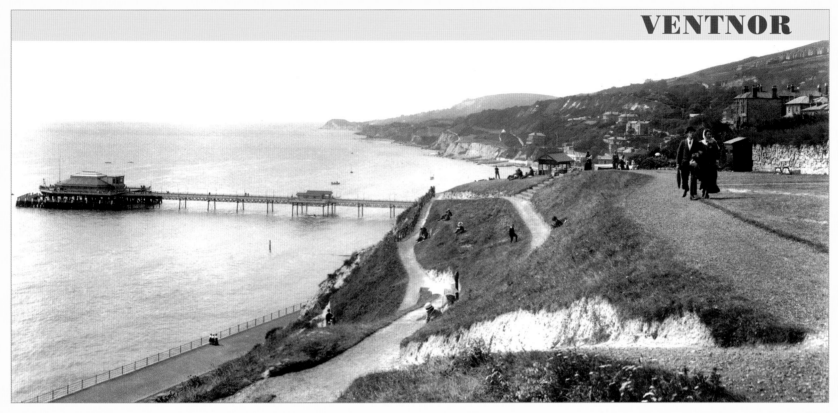

VENTNOR, FROM THE EAST 1913 66156

Ventnor transformed itself from an obscure fishing hamlet to a fashionable watering place during the last half of the 19th century. The eminent physician Sir James Clark made the resort's reputation in Victorian times by comparing its climate to that of Madeira.

Victorian Ventnor became a refuge for consumptives, the kind climate aiding their condition. In a few short years four large homes for sufferers from tuberculosis were established in the resort. The good weather, fresh air and regime of long bracing walks probably did a great deal to alleviate their condition.

The pier was built in 1872 to land visitors directly from the mainland, and to offer steamboat excursions to Bournemouth and Brighton. In 1885 it suffered damage and was rebuilt. At the end of the 19th century it was lengthened. Further substantial rebuilding work was carried out in the early 1950s.

This is the capital of the Undercliff. Its popularity is due to the remarkable salubrity of its climate, and the singular beauty of its situation. 'Forty years ago', says one authority, 'Ventnor contained about half-a-dozen humble cottages; and until the publication of Sir James Clark's work (which, by the way, bore the portentous name of "The Influence of Climate in Prevention and Cure of Disease"), its few inhabitants were nearly all fishermen'. Now we have hotels, churches, shops, cottages and villas in every conceivable style and every outrageous shape. From the Esplanade there extends a fine pier, erected by the Local Board in 1887, and from which steamboat excursions may be made to Bournemouth and Brighton. The Downs above Ventnor can be reached by a road leading from near the railway station. Ventnor is essentially a place that has been made by doctors, and nothing can be more astonishing than the rapidity with which the tiny fishing hamlet was transformed into a fashionable resort.

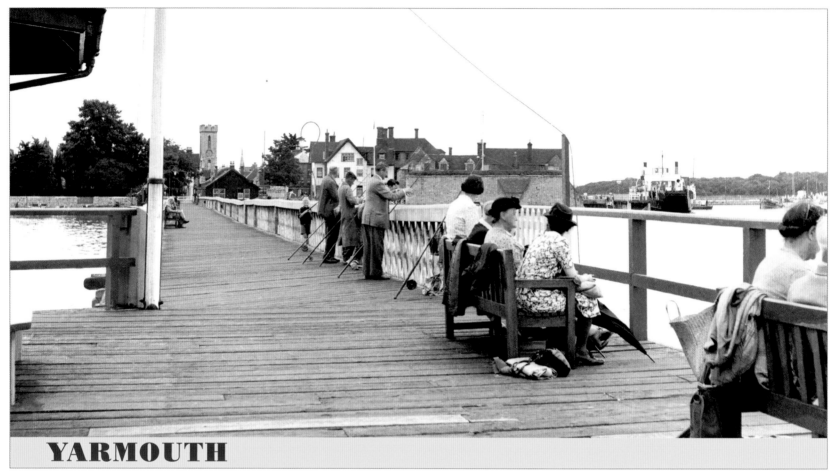

YARMOUTH

YARMOUTH, THE PIER C1955 Y4002

Yarmouth Pier opened in 1876, and was built out of wood, a common material for the less elaborate piers. It was 700ft long. Ferries used to call, but from the 1930s they moved to a nearby slipway – the ferry to Lymington can be seen in the photograph. During the summer season the steamer service from Ryde to Bournemouth called. Excursions bring visitors to this day, with the pier benefiting from an extensive rebuild in the 1980s and 1990s.

This delightful little watering-place is situated at the mouth of the Yar, opposite Hurst Castle. It is more sheltered than Ryde from the keen east winds, and is, at the same time, less exposed than Ventnor to the glaring, burning sun. It possesses a fine natural harbour, and is the head-quarters of the Solent Yacht club.

SOUTHAMPTON

SOUTHAMPTON, THE ROYAL PIER PAVILION 1908   60415

The name comes from the fact that the official opening was carried out in 1833 by the Duchess of Kent and Princess, later Queen, Victoria. Developments in the early 1890s included a pavilion, making the structure a true seaside pier. Yet apart from the 1937 gatehouse – once a popular dance hall but currently closed – it is now hard to tell there was once a pier here, despite its being Britain's second oldest.

The town stands on a wide, V-shaped tract of land projecting into that estuary of the Channel known as Southampton Water, which is navigable by the largest vessels afloat at all states of the tide. Into this estuary, on the east of the town, runs the small but charming Itchen, the river of Izaak Walton. Southampton Water is a landlocked arm of the sea about eight miles long, and averaging one mile in width. The fairway is wide enough for all purposes; and the depth of water is such that the Great Eastern, the largest vessel ever built (680 feet by 83 feet) could swing at anchor close to Southampton. The most remarkable feature of Southampton is, of course, the docks, which cover altogether something like 230 acres, including the quay space.

BOSCOMBE

Above: BOSCOMBE, THE PIER 1931   84887

Above: BOSCOMBE, ON THE PROMENADE C1960   B151028XP

Above: BOSCOMBE, THE PIER C1955   B151013

BOSCOMBE, THE PIER 1908 61191

The pier opened on 28 July 1889, three years after the Boscombe Pier Company had been formed. The local council took over the structure in 1904, at which time Southbourne pier (1888–1907) still existed, a pier they had declined to purchase. In 1905 the council erected both entrance and pierhead buildings.

The pierhead was restored using high alumna concrete in 1924–25 and 1927. The neck was similarly treated in 1958–60; this time it was rebuilt using a type of reinforced concrete that made wide use of non-ferrous components to help prevent marine corrosion.

The Mermaid Theatre, built at the head in 1961, was never used as a theatre, but was altered initially to provide a roller-skating rink. This theatre is now fenced off, whilst the pier neck stays open. The flat-roofed entrance that replaced the one shown has been critically described as an 'elongated bus shelter'.

BOURNEMOUTH

Worm Attack

Wood was never the most successful building material for structures designed to endure the ravages of the English winter as well as being permanently submerged in seawater.

Bournemouth's first pier was one of the very last wooden piers. Just a few years after it was built by Rennie in 1860, it was attacked by the voracious shipworm. In 1867, thoroughly weakened, the pierhead was broken up by a storm, and the structure was replaced at the end of the 1870s.

The resort's first pier opened in 1861, replacing an earlier wooden jetty. Badly storm-damaged, its remains were pulled down. A temporary facility was provided for the 1877 season, and the replacement pier – designed by Eugenius Birch – opened three years later, built at a cost of £22,000. Covered shelters and a bandstand were added to the pierhead in 1885.

Further extensions were made in 1894 and 1905; by this time the pier's length had grown to 1,000ft, coincidentally the same length as Bournemouth's earlier pier. The structure contained a lengthy landing stage, popular with steamers travelling along the South Coast. Around 10,000 people landed here one 1901 Bank Holiday weekend.

The attractive entrance building shown in 45213 (left), which included a clock tower, no longer exists. At this part of the pier today is a two-storey octagonal leisure complex, incorporating shops, kiosks, show-bars and a multi-purpose hall. Costing £1.7 million, it opened in 1981. A new concrete pier neck dates from this time.

The pierhead had been reconstructed in 1950, and a concrete substructure was built ten years later to carry a new pier theatre. This explains why the pier in B163071 (below) looks bare. The structure in the foreground is actually Bournemouth Jetty, built in 1850, but pulled down in the 1970s.

From this select sojourn of delicate ease the working elements of society are for the most part banished to villages and cottages inland ...

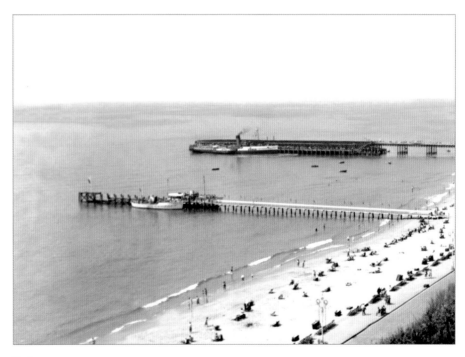

Left: BOURNEMOUTH, THE PIER ENTRANCE 1900   45213

Above: BOURNEMOUTH, THE BEACH AND THE PIERS C1950   B163071

In position and aspect, Bournemouth is unique among English watering-places. From the ever-green valley of the Bourne (whence arose the nucleus of this resort) it stretches for miles in either direction upon the sandy cliffs and pine-clad table-land of a gently curving bay, broken by picturesque chines. Not a few of the residences in Bournemouth are fine mansions standing in extensive and beautiful grounds. From this select sojourn of delicate ease the working elements of society are for the most part banished to villages and cottages inland and out of sight, so that the town itself, instead of being shut in by shabby suburbs, is at most points fringed by pine-woods and moors. Nothing can be more snug and luxuriant than the mouth of the valley, which is here being turned into a long strip of garden, blooming with arbutus, rhododendrons and other choice shrubs.

Above: BOURNEMOUTH, THE PIER APPROACH C1955 B163012

Right: BOURNEMOUTH, THE PIER 1897 40559T

Diving into the Dark

During the early hours of the morning numbers of bathers are attracted to the pier, from the end of which they are enabled to enjoy the luxury of a dive into clear, deep water, from the springboard which is fastened at the landing stage; while in the evening, those who love to see the mantle of the night as it gradually clothes the earth can here watch the last rays of the sun disappearing behind the Dorsetshire hills, and catch a final glimpse of the twilight-shaded bay before returning to their homes.

VICTORIAN GUIDEBOOK

SWANAGE

This is one of the more unusual piers: its 642ft-long neck does not go straight out to sea, but veers rightward. It was built between 1895–97, and was used widely by steamers. However, boat traffic ended in 1966, when the last paddler – the PS 'Embassy' – called.

The pier's turnstile can clearly be seen as part of the promenade, along with iron railings and a traditional pier lamp.

To the right of 40306 (right) we can see what was left of the structure the pier was built to replace. A few piles of the 1859 construction can still be seen: it was used as a diving platform until the 1950s. Swanage's newer pier was taken over by a Trust in 1994; it re-opened in 1998 thanks to Lottery funding.

Above left:
SWANAGE, THE PIER 1897
40306

Right:
SWANAGE, FROM THE
PIER 1897   40305

Below:
SWANAGE, THE PIER 1897
40301

Swanage is unpretending, and its patrons will be those who do not crave for gaiety ...

Since the opening of the railway, Swanage has vastly increased in favour as a watering-place.

It is situated in a beautiful bay, and commands a glorious prospect of down and sea and cliff. Swanage is shut in by a range of chalk hills, about 700ft high, and the coast is indented with numerous wide coves.

The Purbeck peninsula is 12 miles long, and near here, in a deep central valley, lie the mossy ruins of Corfe Castle. The town of Swanage is unpretending, and its patrons will be those who do not crave for gaiety. Beyond its attraction as a family watering-place, the great interest of the neighbourhood is for the geologist.

This pier was known as the Commercial or Pleasure Pier, to distinguish it from the old harbour pier in the town that did not have a leisure function. The pier pictured here dates from 1840; a new passenger landing stage was built in 1888–89 for the Great Western Railway. A Pavilion Theatre was constructed in 1908. The 1,100-seater quickly eclipsed the other smaller theatres around the town, offering a variety of plays all the year round.

The pier was burnt down in 1954. It was 1,050 feet long and was a popular vantage point for watching the steamers come and go. The Edwardian visitor would have paid 2d (old currency) to walk to the end of the pier.

A second Pavilion Theatre opened in 1961, and is still fully functional today. The pier's prime purpose, though, is as a shipping terminal, with ferries travelling to both the Channel Islands and France.

This popular watering-place is very pleasantly situated. The coast here, turning to the south, forms a wide, open bay, shaped in the form of the letter E, the projection in the centre dividing it into two parts, namely, Weymouth Bay and Portland Roads. To the north of this projecting point lies the old town of Weymouth. Its principal feature is the esplanade, which extends along the shore of the peninsula for about a mile, and is lined with elegant houses and defended by a substantial sea-wall. Near the clock tower stands an equestrian statue of George III, erected by the townspeople in 1809.

Above: WEYMOUTH, THE BEACH AND THE PIER C1950   W76102

Right: WEYMOUTH, THE PLEASURE PIER AND THE PAVILION 1909   61589

WEYMOUTH

LYME REGIS

LYME REGIS, VICTORIA PIER 1912   65043

Here we see Cousens and Company's paddle-steamer 'Victoria' arriving from Weymouth. Victoria Pier, constructed out of massive blocks of stone, was built between 1842 and 1848. It used to be known as Crab Head until a visit by Princess Victoria with her mother the Duchess of Kent. On 31 July 1833 the royal yacht 'Emerald' was towed into Lyme Regis by the steam packet HMS 'Messenger' to meet the royal party. They were escorted over the hills by the Earl of Ilchester's Yeomanry, after spending a couple of nights at Melbury House. The Duchess and Princess were met at the Cobb on 2 August 1833 by the Mayor, John Hussey, and passed through a double file of Coast Guards. Then a floating platform and barge took the party and their carriages out to the 'Emerald'. The vessels weighed anchor at three o'clock and set off for Plymouth. This view is north-eastwards to Black Ven (top left), Charmouth and Cain's Folly (centre right).

Cobbler and town crier George Legg, in Silver Street, used to be the agent for steamer operators Cosens and Company. Summer day trips reached beaches, piers and ports from Torquay to Bournemouth.

TEIGNMOUTH

TEIGNMOUTH, THE PARADE 1903   49559

Teignmouth's pier is a plain structure of wood and iron piles, 600ft long. It was opened in 1867, and enlarged 20 years later with the pavilion at the seaward end.

The beach to the east of the pier was exclusive to female bathers who were rolled into the shallows in modesty-preserving bathing machines. They stepped out, gowned from neck to toe in a thick dark material. Clinging to the machine's rope with a cork on the end, they bobbed about in the waves and indulged in a great deal of screaming!

By the 1960s, post-war days with a final blossoming of talent competitions, fashion parades, afternoon tea dances, charity balls and gala dinners had drawn to an end. The best of times were nearly over for the town's pier.

Above: TEIGNMOUTH, THE PIER 1925 78464   *Below:* TEIGNMOUTH, BOWLING GREEN AND THE PIER 1918 68556

PAIGNTON

PAIGNTON, NORTH SANDS FROM THE PIER 1896   38549

The purchasers of Teignmouth pier originally planned to have it moved and reconstructed at Paignton. However, structural difficulties prevented this, so the owner chose instead to build an entirely new pier at Paignton. This opened in 1879, and the pierhead was enlarged two years later.

As a result of the 1881 developments, a billiard-room was erected connecting to the pavilion. However, the pierhead was burnt down by fire in 1919, ushering in a period of decline. Local authority plans to buy the pier were abandoned owing to public opposition.

Modernisation took place in 1980–81, costing a reported £250,000. The shoreward end was widened, making the pier's neck all the same width. New buildings were added. Further development has occurred since the mid 1990s. The pier is now under the ownership of UK Piers Ltd, who also own Skegness pier.

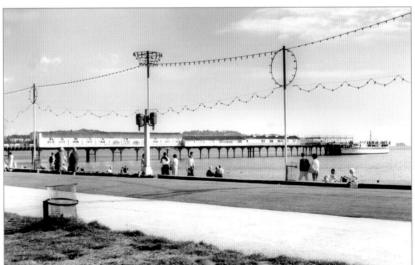

Far left:
PAIGNTON,
THE SANDS AND THE
PIER 1925   78476

Above left:
PAIGNTON, THE PIER
KIOSKS 1889   21529
(DETAIL)

Below left:
PAIGNTON, THE PIER
C1965   P2043

This watering-place may be described as a handsome and extensive suburb of Torquay. Of late it has been greatly improved; a promenade pier has been erected, and the Esplanade – on which there is a band-stand – greatly extended. This charming resort should be visited in the apple blossoming season, for the cider apple is largely cultivated in the neighbourhood, and cider is manufactured on a large scale. Paignton possesses splendid climate and remarkably fine sands. The bathing, too, is excellent.

PLYMOUTH

The Hoe Pier was the last to be designed by Eugenius Birch; he died a couple of months before it opened on 29 March 1884. It was just 480ft long, with facilities which included shops, a clock tower and a landing stage.

A pavilion was added in 1891, reputedly capable of seating 2,000 people. The pier was struggling by the 1930s, though, with a loss of £1,276 recorded in 1934. Bomb damage in 1941 put an end to the controversy over ownership. The War Damage Commission paid the costs of its 1953 demolition.

It is more especially the great national harbour – the principal nursery of our fleet. The Hoe, a slight but commanding elevation partly covered with grass, overlooks Mill Bay and the Sound. The view comprises St Nicholas Island, Mount Edgcumbe, Devonport and Stonehouse. Mount Edgcumbe is undoubtedly the loveliest spot in the immediate vicinity of Plymouth. It is at the extreme end of a promontory, four or five miles long, and has been carefully cultivated into a beautiful and extensive pleasure garden.

Although it does not possess the exotic splendours of Brighton Pier, Looe's modest stone pier played a vital part in the town's economy. Fishermen tied their boats alongside it, conducted their business while squatting on it and enjoying a pipe, and visitors picked a path gingerly along it, to stare at the boats and all the fishing paraphernalia.

During the 19th-century Looe was an important copper port. The quays at Looe were built to serve the Caradon mining district, via a canal and railway, but by the time of these photographs the mines had mostly closed; coal was still imported but there was now room for the fisheries to take over. However, today, Looe is fundamentally a tourist town with just a small fishing fleet. One of the main quays is now a car park and the fish market has been moved to the far end of the quay.

In 56387 (below) we see fishermen posing for the camera, and three well-dressed visitors enjoying the salt atmosphere. In 56390t (right) two fishermen standing on the pier are chatting with a young woman, possibly negotiating the fare for a trip up the river.

Looe is romantically situated in a deep recess, the acclivities above it being hung with gardens, in which the myrtle, hydrangea, and geranium flourish all the year round. It may be described as an old-fashioned town, intersected by narrow lanes, and before the road was made along the water-side, it was approached on the east side by so steep a path that travellers were afraid of being precipitated on to the roofs of the houses. The estuary is confined between abrupt and lofty hills.

Above: LOOE, THE PIER 1906 56390t

Left: LOOE, THE PIER 1906 56387

Opposite: PLYMOUTH, THE HOE PIER 1889 22377

FALMOUTH

The pier was named after the Prince of Wales (later George V), who laid its foundation stone in 1903, two years prior to the pier's formal opening. It is the only seaside pier in Cornwall recognised by the National Piers Society.

Reconstruction took place in 1951; the pier is currently owned by Carrick District Council. Both regular ferry services and excursions depart from here, with different steps performing a similar function to railway station platforms. There are shops at the shoreward end, and shelters at the head.

During the last ten years the beauties of Falmouth have become better known. The landowners and townspeople generally, whilst not paying less attention to shipping – for which Falmouth has always been famous – are rapidly developing the many eligible building sites, and erecting thereon large and commodious houses and charming terraces, which overlook the harbour. The effect is very striking. The old town, quaint and picturesque, is situate on the low ground near the edge of the harbour, and as a matter of course, the streets are very narrow. The new portion of the town lies for the most part on high ground, overlooking the magnificent harbour on the one side and the English Channel on the other. The beautiful Castle Drive forms an esplanade the like of which is not to be met with in the kingdom.

Left: FALMOUTH, PRINCE OF WALES PIER 1927 80095

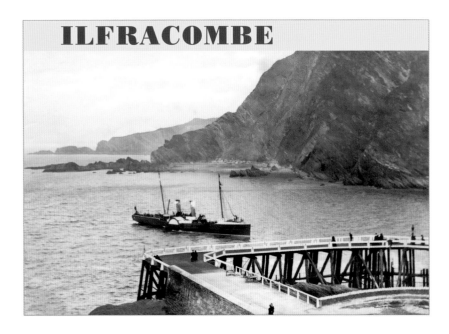

ILFRACOMBE

There had been a pier in the town since 1678; the most recent structure was largely rebuilt out of concrete in 1939–40 and 1952. Demolition of the pier began in September 2000, amidst local objections. It was a steamer pier, and unique in shape.

In 22914 (left) we see the paddle steamer 'Velindra' approaching the pier. She was built at Blackwall in London in 1860. In 1869 the Pockett Company bought her, and used her regularly for the Swansea to Ilfracombe sailing. She was taken out of service and scrapped in 1870.

The parish church contains memorial stones to eight reputed centenarians ...

Ilfracombe is almost encircled by picturesque, verdurous heights of tors, and is remarkable for a peculiar form of coast, rarely to be met with elsewhere. Land and sea together combine in making constant alternations of high, craggy, furze-crowned tors, in some cases almost overhanging the sea below. Perhaps no better testimony to the salubrity of Ilfracombe can be adduced from the fact that the parish church contains memorial stones to eight reputed centenarians.

Top: ILFRACOMBE, THE 'VELINDRA' APPROACHING THE PIER 1890 22914   *Above:* ILFRACOMBE, THE PIER 1899 43117

MINEHEAD

Top: MINEHEAD, THE PIER C1939 M84088

Above: MINEHEAD, THE 'BRISTOL QUEEN' IN THE HARBOUR C1960 M84196

The White Funnel fleet were making regular calls at Minehead by the 1890s and a pier was built in 1901. It had a two-tier terminal which allowed passengers to disembark at any state of the tide. The idea was to attract trippers from Wales, who arrived up to a thousand at a time. At first they were not universally accepted as they tended to lower the tone of the rather genteel resort. Sunday trips were particularly popular, as Welsh pubs were closed then.

In M84088 (above) we see one of Campbell's White Funnel fleet, probably the Britannia, tying up at the pier. The Britannia was built in 1896 and was one of the best known and one of the fastest merchant ships of its day.

Shelters were built on the pier just before the First World War as its 700ft deck was exposed to the elements. There was also a little railway along the deck running a trolley for transporting luggage. The pier was most active in the 1930s. It was demolished in May 1940, when it was thought to be obstructing the line of fire of an aged four pounder gun on the harbour acting as part of Minehead's defences. The gun proved to be inoperable anyway. At the same time thousands of posts were driven into the beach and tank traps erected and the harbour made out of bounds. The concrete foundations of the pier can still be seen at low tide.

CLEVEDON

This pier, 842ft long, was built at a cost of £12,000 using rails torn up from the South West Railway, and opened in 1867. Some think it the most elegant pier in the country. It was hoped that paddle steamers would call on a main route from London to South Wales, but the pier had to rely on trips to places nearer home like Minehead, Ilfracombe and Chepstow.

The wooden pierhead was replaced by a cast iron structure in 1892, and an ornate pavilion was added two years later. Two spans collapsed during routine testing in 1970, after the pier had earlier been declared unsafe. Its future seemed very doubtful.

Yet the pier survived a public enquiry in 1980, with the Environment Secretary calling it an 'exceptionally important building'. A Trust took the pier over, and rebuilding began in 1984. Final work was not completed until 1998, when Sir Charles Elton – great-great-grandson of the original pier company chairman – officiated at the opening ceremony.

This delightful and fashionable watering-place is about fifteen miles from Bristol. As a health resort it occupies a very prominent place, whilst its immunity from the heavy excursion element which affects many seaside towns renders it a veritable haven of rest, commending itself each year more and more to professional men and others from Bristol.

Clevedon is snugly situated with the broad expanse of the Bristol Channel open to its western front. The most popular and fashionable part of the promenade is that known as the Green Beach ... it is provided with an elegant band-stand, a plantation, and a handsome drinking-fountain.

Left: CLEVEDON, THE PIER AND ALEXANDRA GARDENS C1955 C116048

Top: CLEVEDON, THE PIER AND THE PROMENADE 1913 65402

Above: CLEVEDON, THE PIER 1892 31251

WESTON-SUPER-MARE

We cannot hope to enumerate even a tithe of the attractions of this charming resort in the very small space at our disposal ...

The Old or Birnbeck Pier opened in 1867. It consisted of a span joining the shore to a group of rocks called Birnbeck Island, and then a North Jetty projecting on from there. A new pavilion and the Westward Jetty were added in 1898, with a lifeboat jetty following three years later.

The first part of the 20th century was perhaps the heyday of the Birnbeck Pier. Amusements included the Flying Machine or Airships, Helter Skelter, Maze, Bioscope, Waterchute, Switchback, Cake Walk, Zig Zag Slide, Shooting Jungle, Motor and Bicycle Racing, Darts, Ping Pong, Coconut Shies and Aunt Sally. At the entrance, souvenirs, art pottery and seashells were on sale, whilst automatic machines spoke your weight or delivered chocolate bars.

The wooden stakes in the water, just to the right of the main pier, supported the fishermen's stow nets, used to catch sprats and other local fish.

The Westward Jetty (seen on the left of 65354a) lasted until 1923, when it was removed. Island amusements, including the water chute, fell victim to the Grand Pier's amusements in 1933. Today the pier has long been closed, though a land-based Pier View Information Bureau opened in 2000 – the first stage, let us hope, of a restoration programme.

Situated some twenty miles west of Bristol on the Great Western Railway, Weston may be reached within three and a half hours from Paddington. It began to develop into a watering-place in the year 1811. In addition to pure air, Weston has an unlimited supply of pure water from a never-failing spring, owned by the town, which is said to have its source in the Mendip range of hills. There are lovely roads and drives in the immediate neighbourhood, notably through the woods, and around Worlebury Hill. The town possesses one of the most extensive, and certainly one of the safest, bathing beaches in the kingdom; in short, we cannot hope to enumerate even a tithe of the attractions of this charming resort in the very small space at our disposal.

Left: WESTON-SUPER-MARE, BIRNBECK PIER 1913 65354A

Above: WESTON-SUPER-MARE, THE VIEW FROM BIRNBECK PIER 1887 20330

The new Grand Pier Pavilion is seen in 53003 (above), in the year of its opening. 'Pier now open', says one of the notices. Another notice talks of special rates for railway passengers, indicating just how important trains were in bringing visitors to this and other resorts. In the foreground one of the last horse-drawn wagonettes waits for customers. These took tourists out to Cheddar and Burrington.

The pavilion on the pier was destroyed by fire in 1930, but was replaced in 1932-33. This new construction was arguably the biggest single building put up on a pier in the inter-war years, though it never housed a theatre – it was home to a large fun fair instead. In W69079t (right) a few alterations have taken place to the Edwardian entrance to the Grand Pier. The turnstiles have been removed and you can just see the central windbreak and shelter that was built down the length of the pier in the early 1950s. The pier's entrance was rebuilt in 1970.

Opposite:
WESTON-SUPER-MARE,
THE GRAND PIER 1904
53003 (WITH DETAIL
BELOW)

Above:
WESTON-SUPER-MARE,
THE GRAND PIER
C1955 W69079T (WITH
DETAIL BELOW)

Right:
WESTON-SUPER-MARE,
THE SANDS AND THE
PIER 1913 65160

PENARTH

When the 650ft-long pier opened in 1895, it was very plain, with little in the way of leisure facilities other than the refreshment room, florists and weighing-machine seen here. A large shoreward end pavilion was not added until 1927–28, and a concrete landing stage was built at the same time.

A fire on August Bank Holiday Monday 1931 destroyed a small seaward end pavilion, erected in 1907, along with the mid-length shelters and shops. Steamer services ceased in 1981, other than the specials run by Waverley Excursions.

The pier officially re-opened in 1998 after an extensive restoration programme, helped by Lottery funding.

Twenty years ago it might have passed as a dismal, forlorn resort, with the one solitary advantage of a good look-out onto the busy Bristol Channel. Of late, considerable improvements have been made, and it now stands in high favour. Now frequent steamboats ply across the harbour when the tide serves. Penarth stands on a breezy cliff, where fine pleasure grounds have been laid out. The bathing is not to be praised, the water being muddy and the beach very shingly.

Above:
PENARTH, THE PIER 1896
38464T

Above right:
PENARTH, THE PIER 1896
38723

Below right:
PENARTH, THE PIER AND A
HOVERCRAFT 1963   P24184

When it opened in 1898, the pier was the terminus for the Swansea and Mumbles Railway, whose trains can be seen carrying people to their destination on the first official day of pier business. The pier had been promoted by a Mr John Jones Jenkins of the Rhondda and Swansea Bay Railway.

The main addition to the pier itself has been the landing jetty, reconstructed in 1956. However, the landward end buildings, dating from the same time, were replaced in 1998 by a new pavilion, engineered by Masonwood Architectural Consultants who expanded the original conception of David Bateman.

The railway closed in 1959.

Though some good folk of the district still remain faithful to Swansea's rather dreary sands, most resort to The Mumbles, four and a half miles away. This resort is backed by a range of limestone cliffs, and is thus protected from the strong south-westerly breezes. Of course, Swansea Bay has been compared with that of Naples, and it is an interesting fact that in this case, at least, the comparison is really justified. The oyster fishery here is very valuable, and gives employment to 400 men.

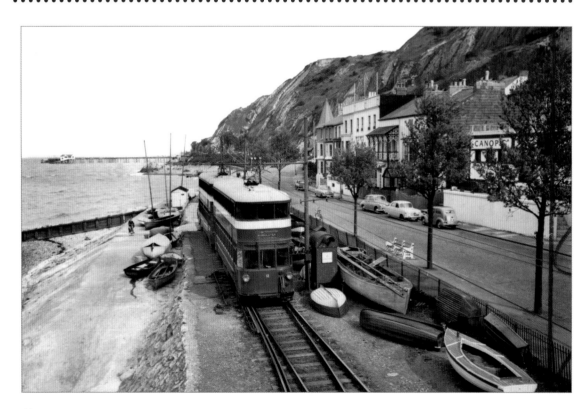

Above: THE MUMBLES, THE PIER 1925   77403
Right: THE MUMBLES, THE SEAFRONT AND THE PIER C1960   M108078

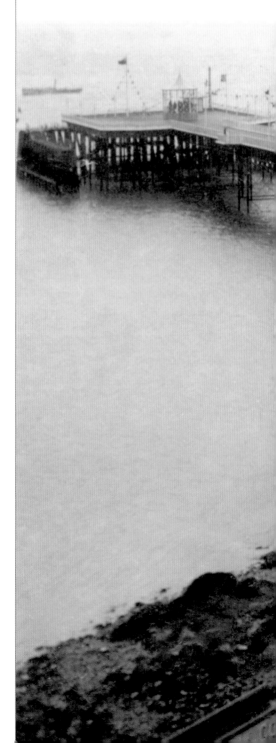

THE MUMBLES

The Royal Victoria Pier steel-arch pier was opened in 1897 in time to commemorate Queen Victoria's Diamond Jubilee. For some time pleasure steamers, in particular the paddle steamers of P & A Campbell, had been calling for day trips to and from such places as Ilfracombe, and a great need for low water landing facilities had been recognised.

Jutting out into the North Bay from Castle Hill, it was a typical Victorian heavy metal creation, of attractive design, with graceful arches, and fine ornamental railings. As well as serving the purpose for which it had been built, accommodating excursion steamers at all states of the tide, it also soon became a popular place for the promenaders, and an excellent place from which to fish with rod and line.

By the time of the official opening, on 9 May, 1899, the pier had been extended, and there was a handsome brass plate, which is now in Tenby Museum, which had already been put in place to record the great occasion, stating that the ceremony had been performed by the Duke and Duchess of York, later to become George V and Queen Mary. This was not so, however. The Royal couple were at Pembroke Dock at the time for the launching of the new Royal Yacht, the 'Victoria and Albert'. The Duke, however, was indisposed, and the future Queen Mary was accompanied to Tenby by the Duke of Connaught, driving from the railway station through beflagged and crowded streets, before walking the carpeted two hundred yards to the pier. HMS 'Renard' was in attendance offshore.

The end of the story is one of the sadder chapters in the town's history. Over the next fifty years little money was spent on maintenance of this marvellous amenity, and by the late 1940s suggestions were being made that the pier should be demolished. It was finally demolished in 1953.

TENBY

Above left:
TENBY, THE PIER 1899 43346

Above right:
ABERYSTWYTH, THE PIER
ENTRANCE 1925 77687

Tenby stands on a tongue of limestone rock, ending in a green promontory, which is crowned by the ruins of the old castle, and is now pleasantly laid out with walks which serve at once as pier and promenade. The sea front of the town stands imposingly displayed on the brow of the cliffs. Both on the north and south sands, which are separated from each other by the Castle Hill and harbour, there are numerous bathing machines; and it should be mentioned that in this respect Tenby is above reproach.

ABERYSTWYTH

I n 1864 the pier was begun but was not complete before part of it was opened on Good Friday 1865 when 7,000 visitors paid their pennies to walk along it; of these 5,000 had come by train, almost doubling the population of Aberystwyth which was then about 6,000. The 'Aberystwyth Observer' commented: 'the vast number of persons that Mr Savin's trains vomited upon the railway platform is proof of the regard in which our town is held'. In September that year the railway also brought large numbers of people to Aberystwyth for the National Eisteddfod.

The pier was 800ft long and about 20ft wide. It was fully opened in October 1865, but in January 1866 a storm destroyed the last 100ft and, despite promises, the company never rebuilt it. It was sold it to Jonathan Pell of the Bell Vue who repaired it in 1872. In 1893 the pier was purchased by Bourne and Grant of the Aberystwyth Improvement Company who were about to build the Hotel Cambria (later the Theological College) and the Cliff Railway. The new Pavilion, large enough for 2,000 people was opened by Prince Edward and Princess Alexandra when they visited Aberystwyth in 1896.

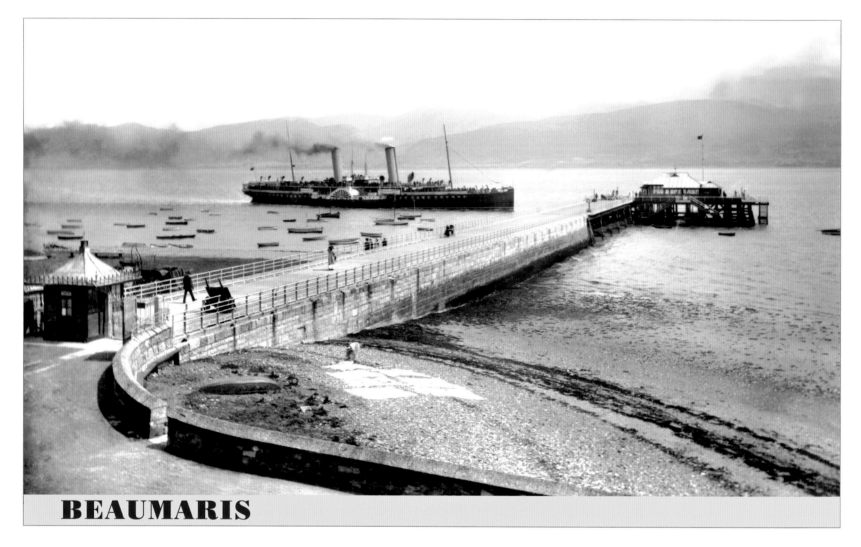

BEAUMARIS

It was easy to get to the very ancient and quiet little town of Beaumaris from Bangor by the steam ferry, which took you across for 6d; or if you started from Liverpool or Caernarfon, you could be landed there direct by the Liverpool steamboats. However, even in Victorian times, most people reached Beaumaris by way of Bangor or Menai Bridge Railway Station, enjoying prospects from the Menai Bridge.

Part piled and part concrete, the pier first opened in 1846, and was then developed in both 1872 and 1895. There used to be a small seaward end pavilion, whereas there are now just shelters here. Today's pier is widely used by anglers, promenaders and those on boat trips.

The North Wales Pier Company's steamers ply the Menai Strait, running sailings between Caernarfon and Liverpool, and calling at Menai Bridge, Bangor, Beaumaris and Llandudno piers.

Above: BEAUMARIS, THE PIER 1911   63294

BANGOR

This has always been an elegant and uncluttered pier, intended primarily as a landing point for steamers and for visitors to enjoy a promenade. It opened in 1896, and was 1,500ft long, with intermittent shelters and kiosks, ornamental lamps and handrail.

The pier was destined to be demolished, but thanks to the sterling work of Bangor City Council, it was restored, and re-opened in 1988. Kiosks sell refreshments and gifts, and the landing stage is still capable of receiving vessels.

At Ballyholme, nearby, may be found a fine beach, and on the other side of the point are bathing-houses, with steps and other conveniences, for both ladies and gentlemen – an excellent substitute for machines, and a feature that might well be imitated by resorts of more pretension.

Above: BANGOR FROM THE AIR 1920 AF2044TR   *Right:* BANGOR, THE PIER 1897 40044

LLANDUDNO

The pier opened to the public in 1877; it replaced a landing jetty that had been pulled down the previous year. A seaward end bandstand was erected at this time, and a shoreward end pavilion was constructed seven years after this.

The pier originally measured an impressive 1,234ft, but this was extended to 2,295ft when an extension opened in 1884 that took the pier past where the Grand Hotel is today, right to the main promenade. A new landing stage was added in 1891, and was reinforced in 1904.

The diversion from the present side entrance at Marine Drive can clearly be seen in 60755 (overleaf). In 1905 a pavilion was built on the pierhead; two years later the pier was damaged by a colliding ship. Major alterations were made to both the pier and landing stage in 1938.

The shoreward end pavilion, empty since 1990, was damaged by fire in 1994, and was later demolished. However, there is still live entertainment in the hotel bar, even if this is a sad come-down from the days when the pier was supposed to have the best seaside orchestra in the country (L71378, overleaf). Nevertheless, Llandudno remains a popular pier.

That's the way to do it!

Mr Punch has a history going back as far as the time of Charles II. His persona is derived from the old Italian commedia dell'arte character 'Pulcinella'. The Codman family have long been involved with him. Since 1860, successive generations have been shouting 'That's the way to do it!' and delighting holidaymakers with their Punch and Judy show at the entrance to Llandudno's pier.

Left: LLANDUDNO, THE PIER 1890   23250

Above: LLANDUDNO, THE PIER C1877   8566

LLANDUDNO, ON THE BEACH 1890 23242P

Above: LLANDUDNO, THE PIER 1908 60755   *Below:* LLANDUDNO, THE PIER ENTRANCE C1955 L71378

There is something unique in the situation of Llandudno. It stands back against the mass of the Great Orme's Head, which shelters it from north winds. The outer bay has a fine sweep, fringed with a long promenade and crescent extending towards the lower and more broken heights of the Little Orme. The Great Orme has now been encircled by a good carriage road. On reaching the summit, the persevering tourist is rewarded by a magnificent bird's-eye view of Llandudno beneath. The bright blue waters of the sea, the hills of Gloddaeth, the majestic ruins and bridges of Conway, all combine to form a prospect of wondrous beauty, which, bounded by the undulating outlines of the mountains, is well worth a pilgrimage to contemplate.

RHOS-ON-SEA

In contrast to the impressively wide and well built promenades seen in Llandudno and elsewhere, the narrow street and the submerged groynes give the impression of a lesser amenity. However, it was still the perfect place for the infirm to enjoy the benefits of the restorative ozone. Rhos, as it is known, is less commercialised than some of its neighbours and remains a gentle mix of resort and rural seaside town.

The pier at Beaumaris, dating from 1896, is quite extraordinarily, the self-same pier as the one erected on the Isle of Man at Douglas in 1869. In 1895 this pier was carefully dismantled and shipped across the bay and then painstakingly pieced together again. It is a tribute to the skills of the early pier designers – in this case John Dixon – that such a feat of engineering could be achieved.

Above: RHOS-ON-SEA, THE SEAFRONT 1921 70794P

Right: RHOS-ON-SEA, THE PIER FROM BRYN EURYN 1921 70802

RHYL

In 1887 the celebrated World Champion diver Tommy Burns was hired to display his daring on Rhyl's pier. Throngs of visitors crowded around the pier waiting for Burns to arrive. When he finally turned up it was obvious to all that he was the worse for wear, having accepted hospitality in a number of Southport's pubs. He was earnestly advised to abandon the show, but like a true professional he resolved to go on. Launching straight into his routine, he dived confidently off the end of the pier and was never seen again.

There were six piers in North Wales by 1900 and Rhyl was the second, opening in 1867. All manner of entertainment was provided in these extravagant buildings from concerts and theatre to minstrels and small 'sketches' and huge numbers would attend. By 1913 this grand building was replaced by a rather less ambitious single storey 'amphitheatre'.

The iron pier at Rhyl cost about £23,000. It was 2,350ft in length, and towards the end of 1891 a grand pavilion was erected at the entrance, capable of seating 2,500 persons. The pier fell into disrepair and was finally demolished in 1977.

Not many years ago there was no town here at all, but merely a few fishermen's huts upon the shore. The sands, which are extensive enough to give the full benefit of ozone to those who avail themselves of its health-giving properties, form an excellent bathing-ground, entirely free from danger. Hence Rhyl has become noted for the number of children that visit it, and these little ones find an inexhaustible fund of pleasure on its beach. Behind the stage in this building is erected one of the largest organs in England, that of the Manchester Exhibition, by Messrs Bishop and Son. Rhyl has exceedingly beautiful inland attractions. Its situation is at the mouth of the River Clwyd, and the Valley of the Clwyd is especially charming.

Left: RHYL, THE BEACH AND THE PIER 1913   65731

Above: RHYL, THE PAVILION AND VICTORIA PIER C1867   2254

The summer exodus of holiday makers to the seaside resorts of Britain made piers a popular and lucrative venture. Colwyn Bay's Victoria Pier has had a chequered history. Opened in 1900, it has been almost destroyed by fire in 1923 and 1933, but rebuilt on both occasions. The Pavilion, originally opened in 1900, could seat 2,500 for its popular entertainment. The 1950s were its swansong and it closed in 1958, reopening as a disco before closing again in 1991.

The pier's past owners, Anne and Mike Paxman, lived on the pier, and did much since taking it over in 1994 to keep the pier in business. They were determined to restore and open all the facilities. Although the majority of the pier neck was reopened, the pavilion remained closed, due to lack of funds. The pier was sold in 2003 to Steve Hunt, who hopes to continue restoring the pier to its former glory.

This watering-place possesses the comparatively rare advantage of being beautifully wooded, the very streets resembling fine boulevards …

From Rhyl to Carnarvon and away down the coast are springing up towns and villages, all in the main catering for the vast army of health and pleasure seekers who annually wend their way to the seaside for days, weeks, or months. Of the fine climate of Colwyn Bay there can be no doubt whatever. The gently-sloping sands extend for a mile or two, and are perfectly safe for children; while the deep water wherein the expert swimmer loves to disport himself is not too far out to be tiresome. The views of sea and coast are very fine, and this watering-place possesses the comparatively rare advantage of being beautifully wooded, the very streets resembling fine boulevards.

COLWYN BAY

Above: COLWYN BAY, THE PIER 1900
46266

Left: COLWYN BAY, THE PIER PAVILION
1900 46268

NEW BRIGHTON

> *If the truth must be told, we fancy that few readers would care to be recommended to New Brighton, except Liverpool people, who already know enough about a place lying so close at hand; but we may mention that this is the chief Mersey bathing-place, which at once gains and loses by its proximity to the great commercial city.*

There were once ten piers on the River Mersey, though only New Brighton was ever regarded as being a seaside pier. It opened in 1867, and included a handsome saloon, refreshment rooms, shelters, a pier orchestra and a tower from where one could watch the ships go by.

After being bought by Wallasey Corporation in 1928, the pier gained a new pavilion – this was deemed to be cheaper than carrying out repairs to the existing structure. Unfortunately, ferry services eventually ceased, causing visitor numbers to drop. The pier closed in 1972, and was pulled down five years later.

Above: NEW BRIGHTON, THE PIER 1900   45165

Left: NEW BRIGHTON, THE 'ROYAL IRIS' APPROACHING THE PIER C1960 N14021

SOUTHPORT PIER
Charges 1868

For every person who shall use the pier for the purpose of walking for exercise, pleasure, or any other purpose, except for embarking or disembarking 6d

For every bath or sedan chair taken on the pier 1s 0d

For every perambulator . . 6d

For every person using the tramway for each single journey and exclusive of luggage 3d

Rates on passengers' luggage. For every trunk, portmanteau, box, parcel, or other package within the description of luggage, not exceeding 14lb in weight 1d

Right: SOUTHPORT, THE PIER FROM THE AIR 1938
AF58487BR

A fine Panoramic View may be had, extending to Black Coombe in Cumberland, Lytham, the Estuary of the Ribble, the Lancashire Hills, the fine Promenade of Southport, the Channel of the Mersey, the Cheshire Hills, the Ormeshead on the Welsh Coast, and to the front of the spectator is the Open Sea ... [it is like] being upon the deck of a ship at anchor in smooth water. You have the breeze and the sea, without the sickness and rolling or pitching of a vessel. It is there the girls get the bloom on their cheeks again, and the pale faces of the town a tinge of the sun.

SOUTHPORT PIER COMPANY PUBLICITY ADVERTISEMENT

SOUTHPORT

Opened back in 1860, this is now the oldest surviving iron pier. Extensions some eight years later took its length to 4,380ft, second only to Southend. Though currently shorter (3,633ft), the pier still retains second place. A tramway carried passengers to the excursion steamers, which landed at the pierhead.

Over the years, much land has been reclaimed, with the present-day pier having to go over a lake, a miniature-golf course and a road before reaching the sea. The pavilion pictured here was destroyed by fire in 1957. Restoration work, supported by the National Lottery, was carried out in 2001. The structure was voted Pier of the Year in 2003.

It claims the title of 'The Montpelier of the North'. We would point out here that there is not a single watering-place in the kingdom that does not arrogate to itself some high-sounding title, the sonorousness of which is, as a rule, in the inverse ratio to the actual importance of the town. Southport is in many respects quite a unique watering-place; to commence with, the sea has long been retreating from it. But Southport has risen to the occasion. It has, so to speak, brought back the receding ocean by means of an extremely long pier; and furthermore, the peculiar pastime of land-yachting may be daily witnessed on the sands during the summer season. Southport is, in reality, a kind of playground for all Lancashire.

Top: SOUTHPORT, THE PIER 1891 28558
Above: SOUTHPORT, THE PIER C1955 S160114
Below: SOUTHPORT, THE PIER TRAIN C1960 S160197

Victorian ladies faced difficulties getting on to Southport pier – their bulky dresses would not fit through the turnstiles. When they complained, the pier company mischievously suggested that they might purchase season tickets permitting them to make use of the much broader gates. The suggestion was not received well!

Lytham, with its sylvan beauty and quiet, tranquil air of contentment and repose, stands on the north bank of the Ribble estuary. It is a very pretty little town, with fine streets, luxuriant foliage on almost every hand, beautiful gardens embowering handsome villas, and a general aspect of unruffled orderliness and neatness of arrangement, such as is seldom seen at other much-frequented resorts. By many this is considered the prettiest watering-place on the Lancashire coast, and it possesses a wealth of foliage and flowers that grow almost down to the water's edge. The beach is well provided with comfortable shelters and seats, and the view from it in clear weather is itself a great attraction. There is a fine sweep of sea, across which may be plainly discerned a great portion of Southport.

LYTHAM

Hardly a stone's thrown from industrial Preston, Lytham's shore and pier were popular destinations for a day-out. At Lytham there were swings, donkeys to ride and Carlton's Pierrots amongst the sandhills. Stretches of beach of the sort depicted here have in recent years been invaded by Spartina grass covering the once golden sand, but in the heyday of the Victorian and Edwardian family holiday the area of dunes was a magical playground for children.

The now-vanished Lytham Pier, built in 1864, was badly storm damaged on October 6th 1903. Two sand barges of 300 tons dragged anchors, drifted and cut the pier in half. Repaired, the pier was destroyed by fire in 1928 and by 1955 demolition was inevitable.

It extended 914ft into the sea and was lighted with electric lamps. At its centre was a handsome and commodious pavilion, in which concerts and other entertainments were given daily.

The white windmill in the background of 67479 (below left) built by Richard Cookson in 1805, was a working corn mill until 1918 when fire damaged it. Horse-drawn carts stopped to collect sacks of flour to transport to Cookson's Bakery and other parts. Part of Lytham's old machinery was transferred to the windmill at Wrea Green. In the late 20th century new sails were fitted by the Gillatt brothers, and the interior now has a pictorial useful history of the mill and many relevant artefacts.

A curious character known as Blind Martin, accompanied by his faithful dog, once used to traipse around Lytham's promenade and beach trundling a harmonium on wheels.

Opposite: LYTHAM, THE PIER AND DONKEYS 1914 67480P

Above left: LYTHAM, FROM THE PIER 1907 59120

Below left: LYTHAM, THE MINSTRELS 1914 67479

St Anne's was always a 'genteel' place, a mood conveyed here by its Victorian pier that was opened in 1885. Ornate arbours and a Moorish-style pavilion provided seats sheltered from the wind and a floral hall hosted shows and concerts. But crowds must have sometimes been a problem, for signs on the lamp posts direct people to 'keep to the right'.

The pier was designed and constructed for the Land & Building Company by Mr A Dowson and opened by Lord Stanley in 1885. In 1899 the Moorish Pavilion was added and the Floral Hall opened in 1910. A three-storeyed pier jetty enabled passengers to embark or disembark at all stages of the tide when the pleasure steamer 'Winnie' made trips to Southport.

Alas, both structures are no more. A fire destroyed the pavilion in 1974, with another blaze wrecking the Floral Hall eight years later. Refurbishment of the rest of the pier occurred in the early 1990s.

This watering-place is situate on the Fylde coast, about midway between Blackpool and Lytham. It stands amidst wild sandhills; and a beautiful town has taken the place of what was formerly a bleak, barren waste. St Anne's is a thoroughly well-planned town; the streets are fine and wide. St Anne's is essentially a resort for families. The children desire no more enjoyable pastime than a ramble among the sandhills on each side of the town. Pleasure steamers from neighbouring resorts periodically call at the pier-head during the summer months to take up and set down passengers. Visitors will also find a well-equipped fleet of sailing boats ever ready to minister to their requirements.

Above: ST ANNE'S, THE PIER C1955 S3064   *Right:* ST ANNE'S, ON THE PIER 1906 53888

BLACKPOOL

In September 1885, Blackpool opened the first fare-paying street tramway in the country to be equipped with electric cars. In 1894 the town pulled off the seaside coupe of the century with the opening of its spectacular Tower. At 518ft high, it was based on the Eiffel Tower in Paris. The Tower was followed with a big wheel, and in 1899 with the opening of the Tower Ballroom, modelled on the Paris Opera House, and famed for its mighty Wurlitzer organ. Blackpool led, while others could only trail along in its footsteps. The town authorities had never been shy of borrowing good ideas from elsewhere. By the beginning of the 20th century it was hardly surprising that Blackpool was the preferred holiday destination for the majority of Lancashire cotton workers. However, it was also drawing in holiday-makers from as far afield as Scotland, the Midlands and the North East.

The pier shown on the left (22881t) opened in 1868 as the South Pier, then changed its name to Blackpool Central when the Victoria (now South) Pier opened. Over the years it became known as the 'People's Pier', specialising in a vast range of amusement activities. A more genteel pier would not have been so covered in advertisements!

A 24-hour din

The 'People's Pier' was the source of much local hostility. When late-night dancing at the shore end on a 'purpose-built' platform was started in 1891 there was much opposition from local residents, concerned that they might lose the patronage of their more middle-class lodgers, and outraged at the constant stamping of drunken revellers and the raucous music of the brass bands. There was no respite in the early morning, for often the bands would start up again as early as 6am, to attract visitors who were arriving on the first excursion trains of the day.

Left: BLACKPOOL, THE SOUTH JETTY ('PEOPLE'S PIER') FROM THE WELLINGTON HOTEL 1890 22881T

The oldest and architecturally the finest of Blackpool's piers, the North Pier opened in 1863 to the designs of Eugenius Birch. An Indian Pavilion and bandstand were added to the pierhead in the 1870s. A pier tramway was installed in 1990, but it is no longer in being. Storm damage in 1997 eventually put paid to the landing and fishing jetty.

At 1,410ft long, it was a pleasure pier without rival in Britain. On Whit Monday crowds came so thick and fast that there was little room for movement. So packed did it become that there was a fear that 'respectable visitors would not go upon the pier during the time that the excursionists were there'. The press of humanity on beach and pier proved the boast of the council that 'Blackpool will not be left behind in any respect'.

Left: BLACKPOOL, THE NORTH PIER 1890   22880

Above: BLACKPOOL, THE NORTH PIER FROM THE SOUTH JETTY 1890   22868

Tradespeople closed their shops; householders made their places of abode gay with flags and bunting; and all classes of the community entered fully into the festive spirit. The only piece of artillery of which Blackpool could then boast – a small twelve pounder cannon – boomed forth in salutation of the advent of the new era of enterprise in Blackpool's history. There was a grand procession, to the martial music of several brass bands, of Freemasons and Friendly Societies, Fishermen and Bathing-machine Attendants, Trades and Labour Societies, Day and Sunday School Children, all of whom were animated with the one object – 'to mark the day as noteworthy'.

THE OPENING OF BLACKPOOL'S FIRST PIER, 1863

At the shoreward end of the Central Pier, the White Pavilion was added during the 1903 entrance work, and later offered dancing. It was demolished in 1966. Facilities on the pier during the Edwardian age included an electric grotto railway, and a shop where glass could be engraved.

In 53855 (left) numerous sign boards are displayed around the entrance, publicising forthcoming events, whilst some of the hoardings advertise popular brands still in existence today, including Oxo and Boots the Chemist.

Left: BLACKPOOL, THE CENTRAL PIER
1906   53855

A rate was raised for the purpose of giving the town's attractions wide advertisement through the medium of handbills and flaring posters; one would hardly think, however, that this was the best way of drawing the most satisfactory class of visitors to 'the finest promenade in England'.

Right: BLACKPOOL,
THE PROMENADE 1890   22875

MORECAMBE

Morecambe's West End Pier is seen below in the year of its opening, 1896.

The pier's most impressive feature was a large pavilion. The structure was extended in 1898, but storms in 1907 and 1927 halved the pier's length to 900ft. In 1917 the pavilion, which had hosted the Viennese Orchestra, was destroyed by fire.

After the Second World War, people flocked to the West End Pier to take part in open-air dancing. Indeed, that was one reason why many people are said to have retired here. However, in 1977 a severe storm left sections of the pier isolated. Rather than pay the £500,000 quoted for repairs, the pier was pulled down the next year.

Right: MORECAMBE, THE WEST END PIER 1899 42867

Above: MORECAMBE, THE WEST END PIER 1896 37387

The Central Pier was the older of the two piers in Morecambe: it opened in 1869 and was enlarged during the following decade. A large pierhead was ideal for the steamers, which used to call in the days before the First World War. Weekly tickets costing 1s (5p) were available for regular visitors.

By the turn of the century a pavilion had been added, and this was subsequently replaced between 1935–36. In 1986, however, the pier closed after the seaward end decking gave way during a disco. Fire later damaged the landward end amusement arcade. Though repair work began in 1991, the pier was eventually pulled down, leaving Morecambe with just a stone jetty.

Right: MORECAMBE, THE CENTRAL PIER 1906 56106

Below: MORECAMBE, THE CENTRAL PIER 1888 21080

Tragedy at Morecambe

On Monday, September 9th, at eleven in the morning, the landing stage at the end of the pleasure pier, which projects far into the sea, was crowded with people, waiting to get on board the steam-boat Express, for an excursion to Blackpool. Part of the floor of this structure, composed of iron gratings supported by iron piers too slender for the unusual weight of such a throng, suddenly broke down beneath them; about fifty men, women and children were thrown into the water. It was not deep enough, on all sides, to drown them immediately, and many of them clung to the undamaged parts of the landing stage, or to the pier, until they could be relieved, there being no high waves. But the fall or shock had probably stunned a few of the weaker, and others had suffered contusions of the limbs, which made them unable to stand, while some endeavouring to reach the steam-boat, got into deep water. An elderly lady, Mrs. Ralph of Carlisle, was taken up drowned, and laid upon the deck of the steam-boat. Several other women, unconscious and almost lifeless when they were lifted out of the water, presently revived … Fractured legs and severe lacerations were suffered by three or four ladies, and there was one case of concussion of the spine, besides many injuries from the effects of the shock, or of the immersion, which might prove more or less serious.

ILLUSTRATED LONDON NEWS, 14 SEPTEMBER 1895

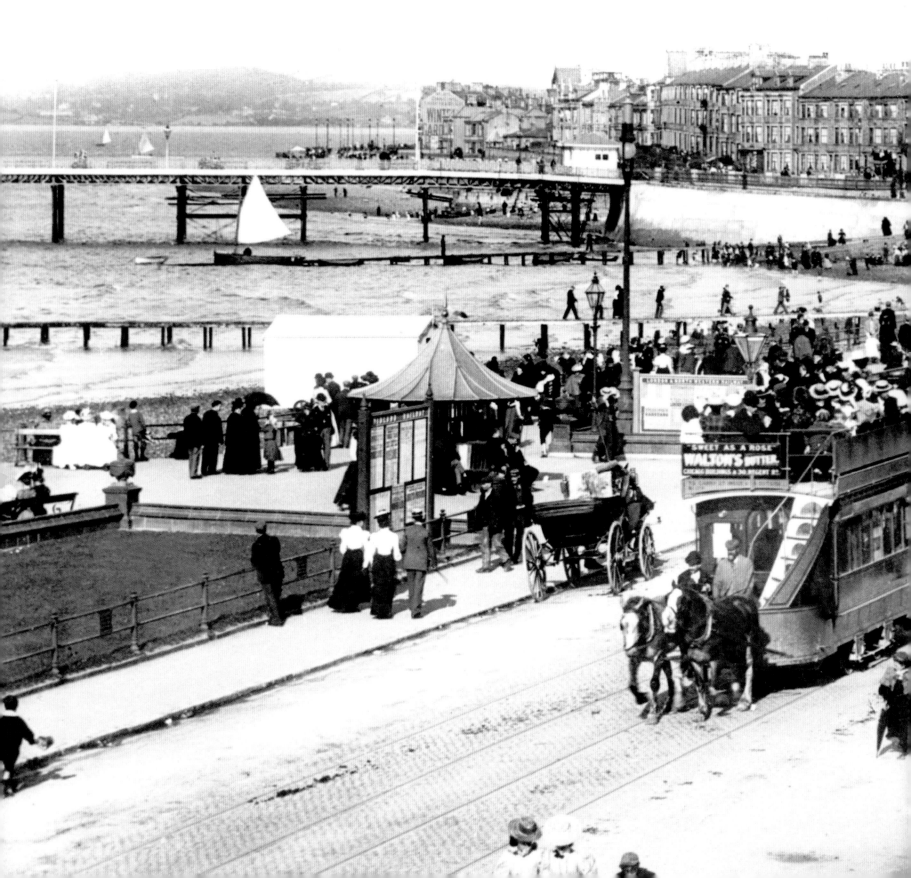

ROTHESAY

Rothesay was the favourite resort for trippers from Glasgow; its esplanade was built in 1870 to cater for those making the legendary pilgrimage 'doon the watter'. The town had a harbour as early as 1752, and a separate pier was finally added in the 1860s. New pier buildings were constructed in 1885, and extensions were completed fourteen years later.

The resort started to decline before 1914, though it remained popular right up until 1939. It has since been unable to regain its former glory – the clock tower seen here was destroyed by fire. Yet boats still call in considerable numbers. The 1899 Victorian toilets gained listed status and were restored in the 1990s.

Here we have the chief town of the County of Bute; it is situated in a well-formed bay, which affords safe anchorage in high wind. A fine esplanade faces the bay, and is laid out with much taste; it commands many beautiful views of Loch Striven. In the centre of the town are the ruins of Rothesay Castle, once a royal residence. On the east side of the island, five miles from Rothesay, is Mount Stewart, the seat of the Marquess of Bute. On leaving Rothesay by steamer, one passes on the left Bannatyne and the bay, and Castle Kames, after which one enters the Kyles of Bute, a sound or strait lying between the north part of Bute and Cowal.

Left: ROTHESAY, THE PIER 1897 39836

Above: ROTHESAY, FROM THE PROMENADE C1955 R61011

Since the demolition of Portobello pier in 1917, this is probably the finest Scottish pier. There has been a structure on the site ever since 1835; the present pier was opened in 1898 by Lord and Lady Malcolm. All the buildings have a distinct Tudor look, and were renovated in 1980–81 at a cost of £175,000.

Though popular with promenaders, this has always been a shipping pier, able to handle two vessels at once if need be. During the Second World War, operations were limited to daylight hours. Continued improvements and re-piling still took place then, unlike at its counterparts south of the border.

This is one of the larger watering-places on the Clyde. The villas at Dunoon extend along the coast to Holy Loch, a short arm of the sea, at the head of which are some fine mountains. Leaving Dunoon, the steamer skirts the shore of Bullwood, where there are numerous fine villas, and shortly afterwards reaches Innellan.

Above: DUNOON, THE PIER C1955 D66045

Right: DUNOON, THE PIER 1901 47423

DUNOON

KIRN

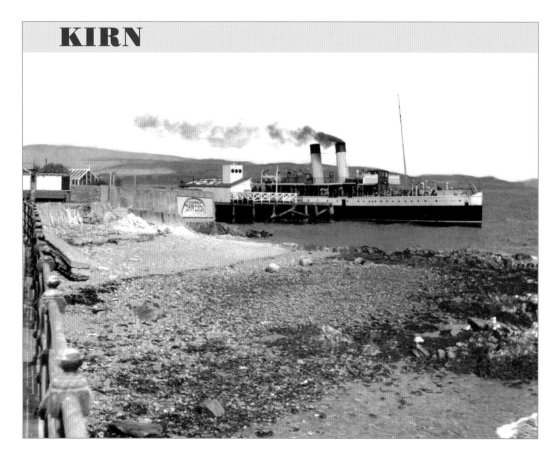

This is one of the older Scottish piers, dating from 1845. But with a change in traffic flow, only four calls a day for the Gourock–Dunoon car ferry were made by the time this photo was taken, along with a few excursion trips. The final visitor was the 'Cowal' in December 1963.

From Gourock a steamer generally crosses direct to Kirn, the quaint little village depicted in this view. Kirn is about 29 miles from Glasgow, and is so situated as to command most extensive views of the Clyde and its adjacent lochs.

Although this little place may be, to some extent, overshadowed by its more popular and fashionable neighbour, it has hotels of its own, and bids fair to become altogether independent of Dunoon. Of course, the proximity of the great Clyde watering-place has done much for the delightful little town shown above; but Kirn possesses so many attractions of its own, in a small way, that of late years visitors have taken up their residence in its cosy villas, and use the place as a sort of head-quarters wherefrom to visit the manifold points of interest in the locality.

Top: KIRN, THE PIER C1955 K52046

Above: KIRN, LEAVING THE PIER C1955 K52066

Situated on the Firth of Clyde, this seaside resort looks across the Firth towards Kilcreggan, Loch Long and Dunoon. It is a centre for yachting and for boating trips in the Firth and to the Kyles of Bute. On the cliff side of Gourock is a prehistoric monolith – 'Granny Kempock's Stone', still linked with ancient myths and superstitions. There are splendid views over the resort and estuary from Gourock Golf Club situated high above the town.

Photograph 45975p (below) shows the backs of buildings along Kempock Street. Kempock Place is just in view on the extreme left of the picture. Over to the right is Seaton's Temperance Hotel, one of several in the town. At this time temperance hotels abounded throughout the UK, but there was in fact little difference between them and private hotels, as neither had liquor licences.

Off Kempock Point, the western boundary of the bay, the steamer 'Comet' was run down by the 'Ayr' in 1825, and 50 passengers were drowned.

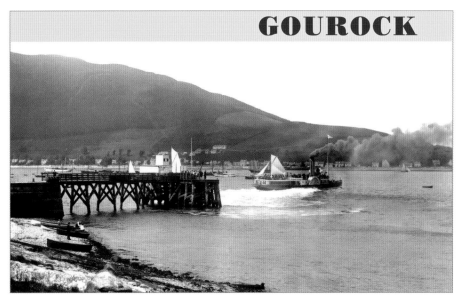

Above: GOUROCK, THE PIER 1897   39821

Below: GOUROCK, FROM THE PIER 1900   45975P

OBAN

Oban actually boasts four piers, one more than Blackpool! Two, however, the South and the Lighthouse piers, are purely industrial structures. The railway pier was built in 1880, and is the headquarters for the Hebridean operations of the shipping company CalMac. To its north can be seen – appropriately enough – Oban North Pier, ten years older and once a steamer pier.

The railway station is situated at the south side of the bay close to the steamboat quay. Many delightful excursions, both by land and sea, may be made from Oban, which grew rapidly in Victorian times into a very popular resort.

This may be fairly called the capital of the West Highlands, and is built along the margin of a semi-circular bay, where the depth of the water affords excellent harbourage for vessels of any class. Being of comparatively recent origin, the streets and buildings have a clean, modern aspect. The latter consist of various churches, banks, and a court-house. A marine parade was formed along the shore in front of the Alexander and Great Western Hotels, and on the heights above numerous villas have been built.

Left: OBAN, THE RAILWAY STATION AND THE NEW PIER C1900   04001

Above: OBAN, SHIPS IN THE BAY 1901   47508

BANGOR

The North Pier at Bangor was used by steamers bound for Belfast; it featured flagpoles, wrought iron seats and gas lamps. Daily sailings ran until 1916, though excursions continued after that date. Some services to Stranraer were diverted to start from here instead of Larne in the 1930s.

The pier was demolished in 1980, when it was described as 'an impressive feat of engineering' by a local paper.

Above left: BANGOR, THE PIER 1897   40240
Below left: BANGOR, THE PIER 1897   40239
Opposite: BANGOR, THE PIER AND THE ESPLANADE 1897   40241

Index

FREE PRINT OF YOUR CHOICE

Mounted Print
Overall size 14 x 11 inches (355 x 280mm)

CHOOSE A PHOTOGRAPH FROM THIS BOOK

Choose any Frith photograph in this book.

Simply complete the voucher opposite and return it with your remittance for £3.50
(to cover postage and handling) and we will print the photograph of your choice in SEPIA
(size 11 x 8 inches) and supply it in a cream mount with a burgundy rule line
(overall size 14 x 11 inches).

Offer valid for delivery to UK addresses only.

PLUS: **Order additional Mounted Prints at HALF PRICE - £8.50 each** (normally £17.00)
If you would like to order more Frith prints from this book, possibly as gifts for friends and
family, you can buy them at half price (with no additional postage and handling costs).

PLUS: **Have your Mounted Prints framed**
For an extra £14.95 per print you can have your mounted print(s) framed in an elegant
polished wood and gilt moulding, overall size 16 x 13 inches
(no additional postage and handling required).

IMPORTANT!

These special prices are only available if you use this form to order.

You must use the ORIGINAL VOUCHER on this page (no copies permitted).

We can only despatch to one UK address.

This offer cannot be combined with any other offer.

Send completed voucher form to:
The Francis Frith Collection, Frith's Barn, Teffont, Salisbury, Wiltshire SP3 5QP

Voucher for *FREE* and Reduced Price *Frith Prints*

*Please do not photocopy this voucher. Only the original is valid,
so please fill it in, cut it out and return it to us with your order.*

Picture ref no	Page no	Qty	Mounted @ £8.50	Framed + £17.00	Total Cost £
		1	Free of charge*	£	£
			£8.50	£	£
			£8.50	£	£
			£8.50	£	£
			£8.50	£	£
			£8.50	£	£

*Please allow 28 days
for delivery.
Offer available to one
UK address only*

* Post & handling		£3.50
Total Order Cost		£

Title of this book. . . . .

I enclose a cheque/postal order for £ . . . .
made payable to 'The Francis Frith Collection'

OR please debit my Mastercard / Visa / Maestro card,
details below

Card Number

Issue No (Maestro only)   Valid from (Maestro)

Expires   Signature

Name Mr/Mrs/Ms . . . .

Address . . . .

. . . .

. . . .

. . . . Postcode . . . .

Daytime Tel No . . . .

Email . . . .

ISBN 0-7537-1441-8   Valid to 31/12/09

Free Print - see overleaf

Can you help us with information about any of the Frith photographs in this book?

We are gradually compiling an historical record for each of the photographs in the Frith archive. It is always fascinating to find out the names of the people shown in the pictures, as well as insights into the shops, buildings and other features depicted.

If you recognize anyone in the photographs in this book, or if you have information not already included in the author's caption, visit the Frith website at www.francisfrith.com and add your memories.

Our production team

Frith books are produced by a small dedicated team at offices in the converted Grade II listed 18th-century barn at Teffont near Salisbury, illustrated above. Most have worked with The Francis Frith Collection for many years. All have in common one quality: they have a passion for The Francis Frith Collection. The team is constantly expanding, but currently includes:

Paul Baron, Jason Buck, John Buck, Jenny Coles, Heather Crisp, David Davies, Natalie Davis, Louis du Mont, Isobel Hall, Chris Hardwick, Neil Harvey, Julian Hight, Peter Horne, James Kinnear, Karen Kinnear, Tina Leary, Stuart Login, Sue Molloy, Sarah Roberts, Kate Rotondetto, Eliza Sackett, Terence Sackett, Sandra Sampson, Adrian Sanders, Sandra Sanger, Julia Skinner, Lewis Taylor, Will Tunnicliffe, David Turner and Ricky Williams.